SIGNALS

CONTINUOUS AND DISCRETE

First Edition

By Iyad Obeid
Temple University

Bassim Hamadeh, CEO and Publisher
Michael Simpson, Vice President of Acquisitions
Jamie Giganti, Managing Editor
Jess Busch, Graphic Design Supervisor
Melissa Barcomb, Acquisitions Editor
Sarah Wheeler, Senior Project Editor
Stephanie Sandler, Licensing Associate

First published in the United States of America in 2014 by Cognella, Inc.

Cover image: Copyright © 2013 by Depositphotos Inc./Giuseppe Ramos.

Printed in the United States of America

ISBN: 978-1-62131-998-6 (pbk)/ 978-1-62661-000-2 (br)

www.cognella.com   800-200-3908

For Jodi, Oliver, and Soraya.

CONTENTS

CHAPTER 3: FOURIER TRANSFORM

CHAPTER 4: SYSTEMS

CHAPTER 5: CONVOLUTION

CHAPTER 6: DISCRETE-TIME SIGNALS AND SYSTEMS

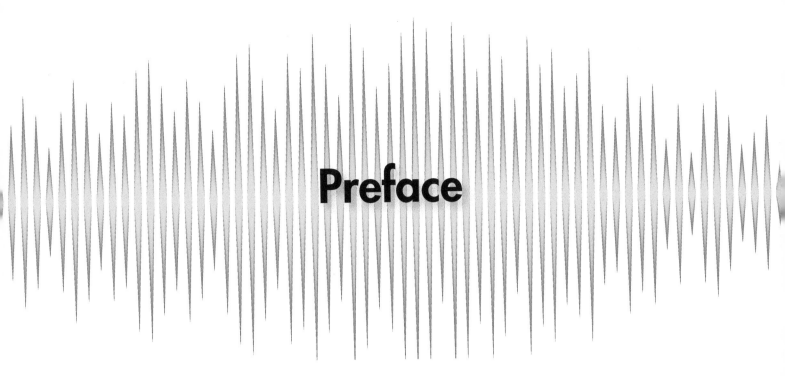

Preface

I never set out to write a book. When I first started teaching Signals at Temple in 2008, I inherited a rich set of course materials built around a great textbook, and things went smoothly for the most part. However as I taught the class over the course of a few semesters, I started to think of ways of streamlining the material. I wondered how much of the material was critical and how much was secondary. And I started experimenting with how to organize the material in order to make it as logical as possible to assimilate all this knowledge in a single semester.

Eventually, as things started to take shape, I started to type up my lecture notes. At first, I did this mostly for myself so that I could remember how I taught things from one semester to the next. Then I'd find myself typing up notes that would clarify some of the more common misconceptions shared by the students. Between these clarifications and the lecture notes, it wasn't long before I had a few hundred pages on my hands. Without ever having intended to, I'd written a book.

The presentation of this material has been largely shaped by my students, in that this has been my attempt to communicate the course concepts to them specifically. In that sense, this book is as much a product of their quest to learn Signals as it is my writing. I suppose that if I'd taught different students, the book would read differently, but either way, I'm optimistic that my text will be valuable to students and professionals from many walks of life.

In writing this book, I've tried to keep the focus on concepts and intuition. I've also eliminated a handful of topics that, while interesting, tend to distract students from the core essentials of the material. I've found that this seems to be an appropriate strategy for an introductory textbook.

I hope you enjoy this book, but more importantly I hope you enjoy Signals and Systems. Feel free to send any feedback my way.

Iyad Obeid, PhD
Philadelphia, July 2013

Chapter 1

Introduction

Section 1.1 What Is Signal Processing?

Signal processing is one of the core disciplines of electrical engineering. In your other electrical engineering courses, you will spend a lot of time learning how various circuits and electronic devices function and how to analyze their behavior. But have you ever stopped to think about *why* you need to learn about resistors and capacitors and all those other hunks of metal and semiconductor? The answer is that electrical signals can convey *information*. Electrical circuits are the best way we have available to encode, store, interpret, and transmit information. Therefore it stands to reason that any electrical engineer worth her salt must not only learn about the tools used to communicate information, but also the very nature of that information itself. And that is why you are about to study Signals and Systems. You should also know that the concepts we are about to learn are valued by engineers in every discipline: mechanical, biomedical, civil, aeronautical, etc. The principles of how to analyze and manipulate information are universally important.

1.1.1 What Is a Signal?

Whenever we as engineers speak of signals, we actually mean "information." While a signal could of course be represented by a circuit voltage, it could also be just about anything. For example, the daily closing price of the Dow Jones Industrial Average or daily water temperatures in the North Atlantic are both examples of signals that vary as a function of time. We can also have signals that vary with respect to parameters other than time. For example, a topographic map displays altitude as a function of location. Anything that represents information is a signal. When you speak to a friend, your vocal cords vibrate the air in a specific pattern; the air pressure encodes the information you are trying to communicate. When that signal reaches your friend's ears, his cochleas convert the signal from air pressure into neural spike patterns. So even the behavior of the billions of neurons in your brain is encoding and processing information. If you as an engineer are interested in capturing or manipulating any of this information, you will need to learn signal processing.

1.1.2 What Is a System?

The other half of this course focuses on "Systems." If a signal is a representation of information, then a system is a means of manipulating or analyzing that information. A system might be a computer program that analyzes the Dow Jones Average to determine whether or not it's a good time to buy stocks. Or a system might be a notch filter that removes 60Hz noise from a biomedical signal such as an electroencephalogram. A system might even be devised that could analyze spike information from hundreds of neurons and try to discern what they are thinking about. Learning about systems goes hand in hand with learning about signals; there is no point in learning about the nature of information if you don't also learn about how to use it to do something useful.

Think about the MP3 player tucked in your pocket. Did you know that signal processing engineers had to come up with a special technique for compressing the file sizes of digitized music? Not only did those engineers have to understand how systems can be built that remove information in order to reduce file size, but they also had to know *which* information should be removed. In fact, those engineers studied human hearing and discovered which frequencies of sound matter least to our subjective enjoyment of music; those frequencies can therefore be removed from the audio signal to reduce the file size. Your MP3 player also uses signal processing to take the digitized audio file and reconstruct it into a voltage signal that can be sent to your headphones.

Does your cell phone have a voice-dialing feature? There again is a real-world example of signal processing. The "signal" is the voice utterance you generate to dial a number, and the "system" is the voice-recognition algorithm that has been coded into a DSP in your phone. Not bad for a phone that your carrier might have actually given you for free!

In short, signal processing is an essential part of electrical engineering. Whether you need to transmit digital information wirelessly over a carrier frequency, remove noise from a biomedical recording, or write a computer program that combines information from multiple noisy sensors, you will need to understand the mathematics behind signals and systems. The goal of this book is to emphasize *concepts*, which means that you will learn how to think about signals intuitively instead of blindly attacking every little problem with calculus. This isn't to say that mathematical rigor isn't very important. It is. But you will be a better engineer if you learn *why* signals work the way they do instead of just how to manipulate signals in a handful of select scenarios.

1.1.3 Organization of Chapter 1

The remainder of this chapter is focused on a quick overview of mathematical topics you will need to know before starting your study of signal processing. This includes a brief review of complex numbers and of cosines. This book is written with the expectation that students have mastered these topics in an earlier course; the information is presented here just as a helpful refresher. We will be leaning heavily on these concepts to build our signal processing tools, so it's worth taking the time to make sure you understand them inside and out. We close the chapter by considering how we might represent the same information by describing it either as a function of time or as a function of frequency.

Section 1.2 Complex Numbers

Complex numbers are special numbers that have both a real and imaginary part. Imaginary numbers are multiples of $j = \sqrt{-1}$. Note that mathematicians often refer to $\sqrt{-1}$ as i, but as engineers, we've already reserved that variable name for current. To avoid confusion, we use j.

Generally speaking, a complex number can be expressed in either *polar* or *rectangular* form, as shown in Figure 1-1.

$$x = a + jb \quad \text{Rectangular Form}$$
$$x = Me^{j\theta} \quad \text{Polar Form}$$

Converting between the two forms is relatively trivial

$$M = \sqrt{a^2 + b^2}$$

$$\theta = \tan^{-1}(b/a) \qquad \text{[1-1]}$$

and

$$a = M\cos(\theta)$$

$$b = M\sin(\theta) \qquad \text{[1-2]}$$

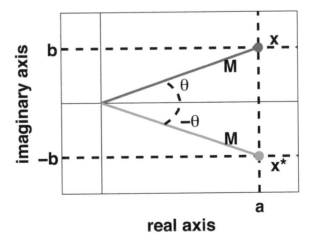

Figure 1-1: Graphical representation of a complex number x. Every complex number can be expressed in rectangular form ($a + jb$) or polar form ($Me^{j\theta}$). The complex conjugate x^* is shown in green. Note that x^* has the same real part as x but the opposite imaginary part.

Finally, you might also remember that every complex number x has a *complex conjugate* x^* defined equivalently as either

$$x^* = a - jb \tag{1-3}$$

or

$$x^* = M\,e^{-j\theta} \tag{1-4}$$

Example 1-1

Convert the complex number $x = 2\,e^{-j\pi/3}$ into rectangular form.

Solution 1-1

The number x is given in polar form. By inspection, its magnitude is 2 and its phase is $-\pi/3$. Or more mathematically,

$$M = |x| = 2$$

$$\theta = \angle x = -\pi/3 \tag{1-5}$$

We can convert x from polar to rectangular form by using Equation [1-2]

$$a = M\cos(\theta) = 2\cos\left(-\frac{\pi}{3}\right) = 1$$
$$b = M\sin(\theta) = 2\sin\left(-\frac{\pi}{3}\right) = -\sqrt{3} \tag{1-6}$$

Therefore $x = 1 - j\sqrt{3}$. This means that x has a *real* part equal to 1 and an *imaginary* part equal to $-\sqrt{3}$. Mathematically stated, this is:

$$Re(x) = 1$$

$$Im(x) = -\sqrt{3} \tag{1-7}$$

Section 1.3 Complex Identities

The formulas of the previous section can be used to derive a handful of useful identities. We will make considerable use of these so it's worth just memorizing them.

$$1 = e^{j0}$$

$$j = e^{\frac{j\pi}{2}}$$

$$-1 = e^{j\pi}$$

$$-j = e^{\frac{j3\pi}{2}} = e^{-\frac{j\pi}{2}} \tag{1-8}$$

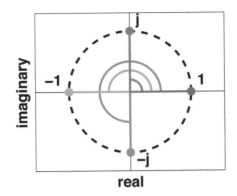

Figure 1-2: Four common complex numbers. [blue] x = 1, phase angle = 0 radians; [red] x = j, phase angle = π / 2 radians; [green] x = −1, phase angle = π radians; [purple] x = −j, phase angle = 3π / 2 radians or equivalently −π / 2 radians. The dotted black line represents the unit circle.

Figure 1-2 shows a graphical representation of Equation [1-8]. Each number is on the unit circle, and therefore (by definition) their magnitudes are all $|x| = 1$.

There is another identity worth memorizing. We will sometimes encounter the number $\frac{1}{j}$. We can get the j out of the denominator by multiplying numerator and denominator by j. Therefore

$$\frac{1}{j} = \frac{1}{j} \cdot \frac{j}{j} = -j = e^{-j\frac{\pi}{2}}$$

Section 1.4 Cosines and Frequency

The most common way to think about signals is as a function of time. In other words, if you were to plot some signal (e.g., an audio signal), you would put time on the x-axis. In signal processing, however, we commonly think of signals as a function of frequency. A plot of a signal as a function of frequency would tell you how much information is contained in the signal at a given frequency. One of the fundamental theorems that we will address later is that it is entirely equivalent to represent information as a function of time or of frequency—both representations tell you exactly the same information. But as we will learn, there are some important advantages we gain by thinking about signals in the frequency domain.

Before we get into those details, let's refresh our memory about what we mean by the term "frequency." We start with a basic cosine, which we can express in one of two equivalent ways:

$$x(t) = \cos(2\pi ft) = \cos(\omega t) \qquad [1\text{-}9]$$

We say that the signal $x(t)$ has a frequency of f Hertz (Hz) or ω (omega) radians per second (rads/sec). Hertz measures "cycles per second," meaning that if your cosine repeats itself three times in one second, then its frequency is 3 cycles/sec or 3 Hz. Equivalently, we recall that each cycle of a cosine comprises 2π radians. Therefore if the frequency is 3 Hz, then in one second we experience $3 \times 2\pi = 6\pi$ rads/sec. We can use Equation [1-9] to derive a conversion from Hz to rads/sec:

$$\omega = 2\pi f \qquad [1\text{-}10]$$

You might also recall that the reciprocal of frequency (in Hz) is the "period" T, which is the time (in seconds) required for one cycle. For example, if a cosine's frequency is 5Hz, then it takes 1/5 of a second to complete one cycle.

Example 1-2
Suppose $x(t) = cos(4t)$. Solve for f, ω and T, and plot $x(t)$.

Solution 1-2
By inspection, $\omega = 4$ rads/sec, and therefore $f = \omega / 2\pi = 2 / \pi$ Hz. Furthermore, $T = 1/f = \pi/2$ s. We can plot our result easily enough, as shown in Figure 1-3.

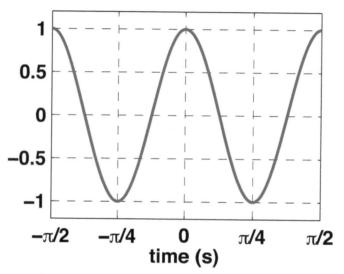

Figure 1-3: Sample cosine signal $x(t) = cos(4t)$ showing even symmetry.

Figure 1-3 illustrates a nice property of the cosine function, which is that it is *even*. This means that it is symmetric across the y-axis, or equivalently $f(-x) = f(x)$. There are also *odd* functions, such as the sine wave, which are symmetric across the line $y = x$ or equivalently $f(-x) = -f(x)$, as shown in Figure 1-4.

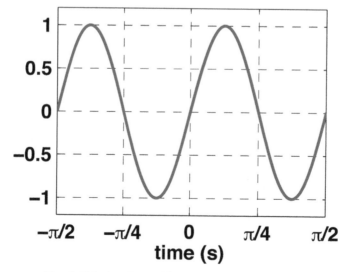

Figure 1-4: Sample sine wave $x(t) = sin(4t)$ showing odd symmetry.

Example 1-3

Suppose $x(t) = \cos(10\pi t)$. Solve for f, ω, and T, and plot $x(t)$.

Solution 1-3

By inspection, $x(t) = \cos(10\pi t) = \cos(2\pi 5t)$ and therefore $f = 5$ Hz. Since $\omega = 2\pi f$ we find that $\omega = 10\pi$ rads/sec. Finally, $T = 1/f = 0.2$ s. The plot is shown in Figure 1-5.

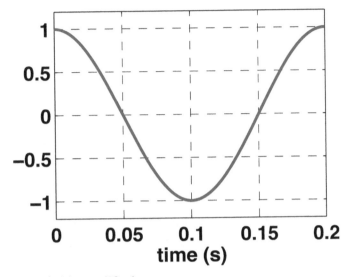

Figure 1-5: Sample cosine signal $x(t) = \cos(10\pi t)$.

Let's put some of these ideas together and consider some basic signals in both the time and frequency domains. Consider Figure 1-6 below, which plots $x(t) = \cos(2\pi t)$. In the time-domain (left) plot, we clearly see the cosine we expect. In the frequency-domain (right) plot, we see that our signal has energy only at $f = 1$ Hz. There is no energy at any other frequency. Therefore both plots tell us the same thing: that our signal is a pure tone cosine at $f = 1$ Hz.

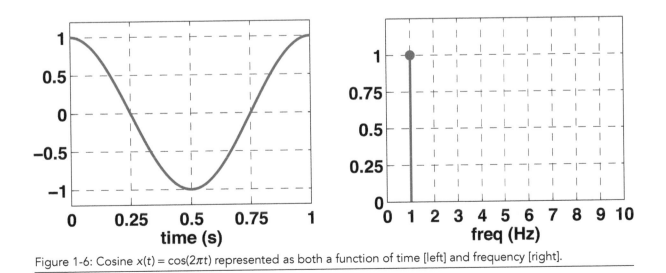

Figure 1-6: Cosine $x(t) = \cos(2\pi t)$ represented as both a function of time [left] and frequency [right].

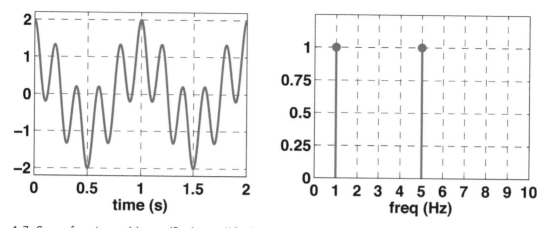

Figure 1-7: Sum of cosines $x(t) = \cos(2\pi t) + \cos(10\pi t)$ represented as both a function of time [left] and frequency [right].

Now let's consider a slightly more complicated signal, $x(t) = \cos(2\pi t) + \cos(10\pi t)$. Figure 1-7 shows the time- and frequency-domain plots. The time-domain plot shows two components: a slow up-and-down portion that repeats itself once a second (the 1 Hz signal) and a higher frequency signal that repeats itself five times per second (the 5 Hz signal). The frequency-domain plot tells us precisely the same information: our signal contains energy at both 1 Hz and 5 Hz. Importantly, the frequency domain plot tells us that our signal contains equal amounts of signal at 1 Hz and 5 Hz.

Contrast this with Figure 1-8, which plots $x(t) = \cos(2\pi t) + (1/4)\cos(10\pi t)$. In that case, it is evident from both the time- and frequency-domain representations that the signal contains energy at both 1 Hz and 5 Hz, but that there is only (1/4) as much signal at 5 Hz as there is at 1 Hz.

In these basic examples, the goal has been to convince you that information (signals) can be represented equivalently as functions of both time and frequency. In other words, if you know the time representation of a signal, you can infer its frequency representation, and vice versa. In these cases, it has been relatively straightforward to conceive what the frequency-domain representation will look like. However, as we come to consider increasingly sophisticated signals, we will no longer be able to simply "eyeball" the time-domain plot in order to determine the frequency representation. Instead we will have to develop some mathematics that will do the transformation for us. This will be the subject of Chapter 2.

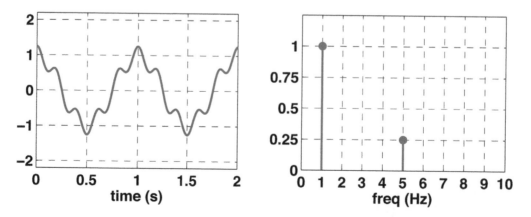

Figure 1-8: Sum of cosines $x(t) = \cos(2\pi t) + (1/4)\cos(10\pi t)$ represented as both a function of time [left] and frequency [right].

Section 1.5 Redefining the Cosine

1.5.1 General Form of the Cosine

The most general form of a cosine is

$$x(t) = K \cos(wt + \phi) \qquad [1\text{-}11]$$

where K is the amplitude, ω is the frequency (in radians per second), and ϕ is the phase shift (in radians). We already learned in Section 1.4 that $\omega = 2\pi f = 2\pi / T$ where T is the period (in seconds). The phase shift ϕ tells us how many radians the cosine is shifted to the left, recalling that one period of a cosine always corresponds to 2π radians.

Example 1-4
Plot $x(t) = 4\cos(6t - \pi / 3)$

Solution 1-4
Firstly, the amplitude $K = 4$ tells us that the cosine will oscillate between a maximum value of 4 and a minimum value of -4. Next we consider the frequency: $\omega = 6$ rads/sec, which implies $f = \omega / 2\pi = 3/\pi$ Hz and consequently $T = 1/f = \pi/3$ s. Finally we calculate the phase shift: $\phi = -\pi/3$ radians. The minus sign tells us that the cosine is shifted to the right. To determine how far to the right, we recall that 2π radians always correspond to one period. Therefore $\pi/3$ radians must correspond to a phase shift of $\frac{\pi/3}{2\pi} = 1/6$ of a period. Since each period is $T = \pi/3$ seconds, then the phase shift must be equal to $\frac{1}{6} \times \frac{\pi}{6} = \frac{\pi}{18}$ seconds.

An equivalent (and easier) way of determining the phase shift is just to factor Equation [1-11] as follows:

$$x(t) = K \cos(\omega t + \phi)$$

$$= K \cos\left(\omega \left(t + \frac{\phi}{\omega} \right) \right) \qquad [1\text{-}12]$$

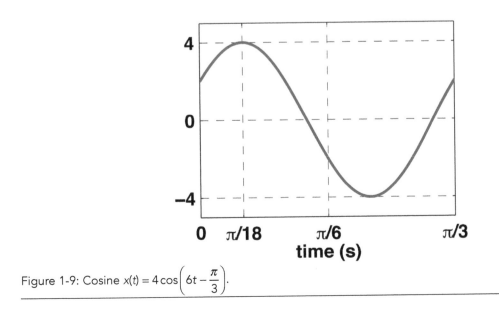

Figure 1-9: Cosine $x(t) = 4\cos\left(6t - \dfrac{\pi}{3}\right)$.

Equation [1-12] shows that the phase shift term shifts the cosine to the right by $\frac{\phi}{\omega} = \frac{\pi/3}{6} = \frac{\pi}{18}$ s.

Combining all these details together yields the plot, shown in Figure 1-9. Note that the period is $T = \pi/3$ s and that the cosine is shifted right by $\pi/18$ s, which corresponds to 1/6th of a period.

Example 1-5

Supposing we now double the period. Plot the new function $x(t)$.

Solution 1-5

Since $T = 2\pi/3$ now, we can determine that $\omega = 2\pi/T = 3$ rads/sec. The phase shift is still $\phi = \pi/3$ radians, which still corresponds to 1/6th of a period. However since the period is now twice as long as before, the time shift will also be doubled to $\pi/9$ s. Alternatively, the time delay can be calculated by the formula $f/w = (\pi/3)/3 = \pi/9$ s. The plot is shown in Figure 1-10.

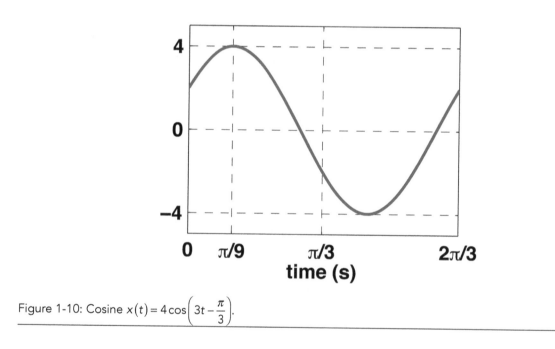

Figure 1-10: Cosine $x(t) = 4\cos\left(3t - \frac{\pi}{3}\right)$.

1.5.2 Euler's Formula

One of the most famous and most useful equations in all of mathematics is Euler's Formula, which states

$$e^{j\theta} = \cos(\theta) + j\sin(\theta) \qquad [1-13]$$

Recalling that cosine is an even function and sine is an odd function, we can also state

$$e^{-j\theta} = \cos(\theta) - j\sin(\theta) \qquad [1-14]$$

By summing Equations [1-13] and [1-14], we see that $e^{j\theta} + e^{-j\theta} = 2\cos(\theta)$, which we can rewrite as:

$$\cos(\theta) = \frac{1}{2}e^{j\theta} + \frac{1}{2}e^{-j\theta} \qquad [1\text{-}15]$$

Along the same lines as Equation [1-15], we can write an equivalent definition for the sine function:

$$\sin(\theta) = \frac{1}{2j}e^{j\theta} - \frac{1}{2j}e^{-j\theta} \qquad [1\text{-}16]$$

Finally, by combining Equations [1-11] and [1-15], we can write an alternate expression for the general cosine form

$$x(t) = K\cos(\omega t + \phi)$$

$$= \frac{K}{2}e^{j(\omega t + \phi)} + \frac{K}{2}e^{-j(\omega t + \phi)}$$

$$= \frac{K}{2}e^{j\phi}e^{j\omega t} + \frac{K}{2}e^{-j\phi}e^{-j\omega t} \qquad [1\text{-}17]$$

1.5.3 Cosine: Complex Exponential Form

By carefully examining the cosine expression in Equation [1-17], we see that the coefficients of the two terms are complex conjugates. In other words, we can use Equation [1-4] to rewrite Equation [1-17] as

$$x(t) = K\cos(\omega t + \phi) = A\, e^{j\omega t} + A^* e^{-j\omega t} \qquad [1\text{-}18]$$

where

$$A = \frac{K}{2}e^{j\phi} \qquad [1\text{-}19]$$

Although it may not look like it, Equation [1-18] is an equivalent way of expressing the general cosine form of Equation [1-11] provided that $|A| = K/2$ and $\angle A = \phi$. We'll see later why it might be preferable to express cosines in this manner. For now, however, we'll focus on a few examples.

Example 1-6
Express $x(t)$ as a cosine when (i) $A = 7$, (ii) $A = 3e^{-j\pi/4}$, and (iii) $A = 5e^{-j\pi/2}$. In all cases, let $\omega = 2$ rads/sec.

Solution 1-6
  i.) $|A| = K/2 = 7$ and $\angle A = \phi = 0$. Therefore $K = 14$ and $x(t) = 14\cos(2t)$.
  ii.) $|A| = K/2 = 3$ and $\angle A = \phi = -\pi/4$. Therefore $K = 6$ and $x(t) = 6\cos(2t - \pi/4)$.
  iii.) $|A| = K/2 = 5$ and $\angle A = \phi = -\pi/2$. Therefore $K = 10$ and $x(t) = 10\cos(2t - \pi/2)$

The plots of the three example cosines are shown in Figure 1-11.

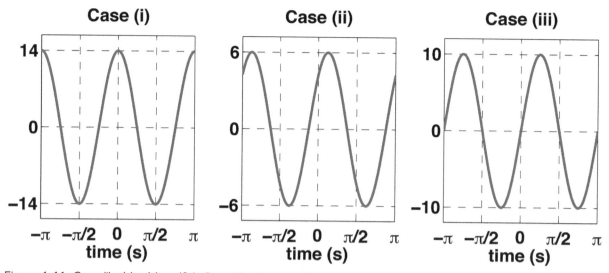

Figure 1-11: Case (i) $x(t) = 14\cos(2t)$, Case (ii) $x(t) = 6\cos(2t - \pi / 4)$, Case (iii) $x(t) = 10\cos(2t - \pi / 2)$. Note that in Case (i), $x(t)$ is a purely even function, whereas in Case (iii) $x(t)$ is a purely odd function. In Case (ii), $x(t)$ is neither even nor odd.

There is an interesting lesson lurking here. In the first case, A is purely real, and the resulting $x(t)$ is a pure cosine, which is an *even* function. Remember that a function is even when it is symmetric across the y-axis, or more precisely when $f(-x) = f(x)$. Conversely, in the last case, A is purely imaginary, because $A = 5e^{-j\pi/2} = -5j$. The resulting signal is $x(t) = 10\cos(2t - \pi / 2) = 10\sin(2t)$, which is an *odd* function. Odd functions are symmetric across the line $y = x$, or equivalently $f(-x) = -f(x)$. In Case (ii), A is neither purely imaginary nor purely real, and correspondingly, its $x(t)$ is neither purely odd nor purely even. This is an important observation and we will come across it again and again in our study of signals.

Section 1.6 Plotting vs. Frequency

We now begin to think about how we can use Equation [1-18] to represent cosines in the frequency domain. As Equation [1-18] suggests, each cosine can be thought of as a complex exponential function that is weighted by its coefficient, namely the complex term A. The term $e^{j\omega t}$ tells us that we are dealing with a cosine of frequency ω. The coefficient A tells us *how much* of that cosine we have. By comparing a few simple examples, we hope to gain some insight as to precisely what the coefficient A denotes.

Example 1-7
Convert the following signals into complex exponential form:
   i.) $x(t) = \cos(t)$
   ii.) $x(t) = \cos(2t)$
   iii.) $x(t) = 2\cos(2t)$
   iv.) $x(t) = \cos(2t - \pi / 3)$
   v.) $x(t) = \cos(2t) + 3\cos(5t - \pi / 8)$

Solution 1-7

Using Equations [1-18] and [1-19] we can find the solutions by inspection:

i.) $K = 1$ and $\phi = 0$ and therefore $A = \frac{1}{2}$ and $x(t) = \frac{1}{2}e^{jt} + \frac{1}{2}e^{-jt}$

ii.) $K = 1$ and $\phi = 0$ and therefore $A = \frac{1}{2}$ and $x(t) = \frac{1}{2}e^{j2t} + \frac{1}{2}e^{-j2t}$

iii.) $K = 2$ and $\phi = 0$ and therefore $A = 1$ and $x(t) = e^{j2t} + e^{-j2t}$

iv.) $K = 1$ and $\phi = -\pi/3$ and therefore $A = \frac{1}{2}e^{-j\frac{\pi}{3}}$ and $x(t) = \frac{1}{2}e^{-j\frac{\pi}{3}}e^{j2t} + \frac{1}{2}e^{j\frac{\pi}{3}}e^{-j2t}$

v.) $K_1 = 1$ and $\phi_1 = 0$ and therefore $A_1 = \frac{1}{2}$. Then, $K_2 = 3$ and $\phi_2 = -\frac{\pi}{8}$ and therefore $A_2 = \frac{3}{2}e^{-j\frac{\pi}{8}}$.

Hence, $x(t) = \left(\frac{1}{2}e^{j2t} + \frac{1}{2}e^{-j2t}\right) + \left(\frac{3}{2}e^{-j\frac{\pi}{8}}e^{j5t} + \frac{3}{2}e^{j\frac{\pi}{8}}e^{-j5t}\right)$

By inspecting these results, we can determine some interesting and useful trends. Recall that the A term tells us how much of a particular cosine we have. Recall also that A is, in general, a complex number, although in some special cases it might simplify to being purely real or purely imaginary. Cases (i) through (iii) are examples of special cases where A is purely real. This is because there is no phase shift in any of those three cases. In other words, it is because $\phi = 0$ rads.

But more importantly, let's compare Cases (i) and (ii). In both cases, $A = 1/2$ even though the signals are at different frequencies. This indicates that in both cases, we have the same "amount" of those cosines. This should make intuitive sense: both cosines have an amplitude of 1 (i.e., $K = 1$) and zero phase shift (i.e., $\phi = 0$). This observation reinforces the earlier assertion that the A coefficient tells us how much cosine we have while the $e^{j\omega t}$ term tells us the frequency of the cosine.

Comparing Cases (ii) and (iii) shows us that increasing the cosine's amplitude means making a corresponding change to the magnitude of A. By the same token, comparing Cases (ii) and (iv) shows us that adding a phase shift to a signal changes only the phase of A, not the magnitude. This is an important point and should make intuitive sense. If a cosine is shifted in time, we still have the same "amount" of that cosine—its amplitude hasn't changed at all. Therefore we expect that the magnitude of A should not change. Instead, the time shift is manifested as adding a phase shift to A.

Case (v) shows that if a signal $x(t)$ is a sum of multiple cosines, we can express that signal as a corresponding sum of weighted exponentials.

The next step is to create plots that display all this information graphically. Plots of a signal's frequency content are valuable because they allow us to quickly assess which frequencies possess the most energy and what the distribution of energy looks like across the frequency spectrum (i.e., all spread out or all bunched up, etc.). Therefore, we wish to plot A versus $j\omega$. Unfortunately, since A is a complex number, it is impossible to accomplish this in a single two-dimensional plot. Instead we must create *two* plots for each signal, one for $|A|$ versus $j\omega$ and another for $\angle A$ vs $j\omega$. These are known as *magnitude* and *phase* plots, respectively. Figure 1-12 shows the resulting plots for the five sample signals above.

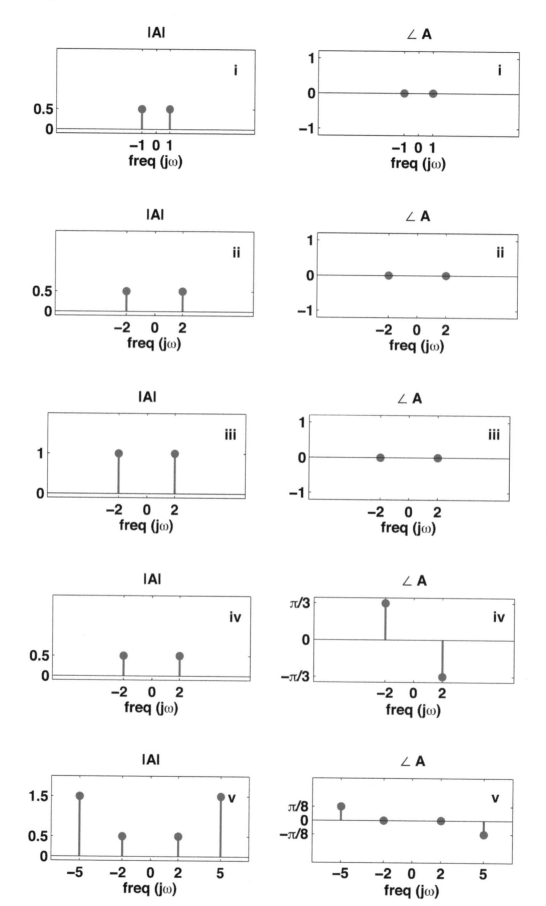

Figure 1-12: Graphical representations of the complex exponentials from the example above. The left column shows the magnitude of A; the right column shows the phase of A.

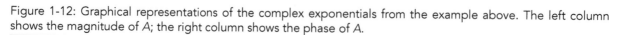

The most important observation of Figure 1-12 is that neither the amplitude nor the phase plot alone is a complete description of the signal in question. Instead, both plots are necessary to uniquely define the signal $x(t)$. By comparing the plots it is easy to quickly understand the frequency content of the various signals as well as the relative strengths of the cosines at each frequency.

Finally, note that all five of the magnitude plots share the important quality of being even, whereas all the phase plots are odd. This is an interesting and important observation. For both the magnitude and phase plots, we see that if we know what the plot looks like for positive values of $j\omega$, then we automatically know what the plot must look like for negative values of $j\omega$. In other words, plots showing the negative values of $j\omega$ are showing redundant information. For this reason, the plots of Figure 1-12 are often shown with only the positive values of $j\omega$ for the sake of simplicity and clarity. Figure 1-13 illustrates how those plots would look.

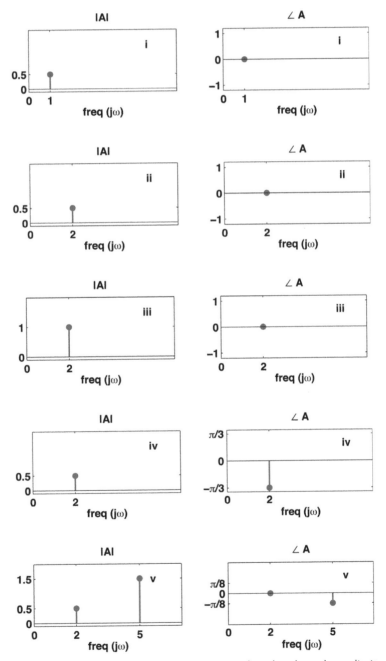

Figure 1-13: The same data are shown as in Figure 1-12 except the plots have been limited to $j\omega > 0$. This plot communicates exactly the same information as Figure 1-12 in a more compact presentation.

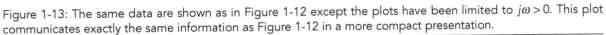

Section 1.7 Summary

A "signal" is any representation of information, and a "system" is a mechanism for operating on a signal. While many signals can be thought of as functions of time, we can also think of those same signals as functions of frequency. Time and frequency representations tell you the exact same information, just in a different way.

This chapter was also concerned with reviewing complex numbers and cosines. At the end of the chapter, we saw how to express cosines as a sum of complex exponential functions. Finally, we saw that in order to plot a cosine function with respect to frequency, it is necessary to create *two* plots, one representing the magnitude of the cosine and one representing the phase.

The most important formulas are summarized here.

$$\cos(\theta) = \frac{1}{2}e^{j\theta} + \frac{1}{2}e^{-j\theta}$$

$$\sin(\theta) = \frac{1}{2j}e^{j\theta} - \frac{1}{2j}e^{-j\theta}$$

$$K\cos(\omega t + \phi) = Ae^{j\omega t} + A^* e^{-j\omega t} \text{ where } A = \frac{K}{2}e^{j\phi}$$

$$\omega = 2\pi f = 2\pi / T$$

$$a + jb = Me^{j\theta} \text{ provided } M = \sqrt{a^2 + b^2}, \theta = \text{atan}(b/a), a = M\cos(\theta), b = M\sin(\theta)$$

$$1 = e^{j0}, \ j = e^{j\frac{\pi}{2}}, \ -1 = e^{j\pi}, \ -j = e^{j\frac{3\pi}{2}}$$

Chapter 2

Fourier Series

Section 2.1 Weighted Sums of Orthonormal Basis Functions

As we learned in Chapter 1, engineers often need to consider how signals contain information at different frequencies. We do this by decomposing signals into weighted sums of cosines. The cosine indicates the particular frequency and the weight indicates the amount of energy at that frequency.

Conceptually, this is very similar to how vectors are described. If you've ever studied vector math, you learned that any vector can be expressed as a weighted sum of unit vectors. For example, in Figure 2-1, you can see that vector $\vec{x}$ is a weighted sum of unit vectors $\hat{\imath}$ and $\hat{\jmath}$.

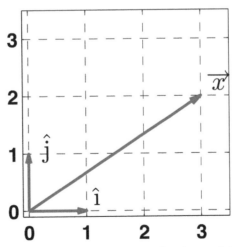

Figure 2-1: Example of a vector $\bar{x}$ that can be expressed as a weighted sum of the orthonormal basis vectors $\hat{\imath}$ and $\hat{\jmath}$.

Specifically, $\vec{x} = 3\hat{i} + 2\hat{j}$. In this example, we can think of $\hat{i}$ and $\hat{j}$ as building blocks that are used to construct vector $\vec{x}$. The weights, 3 and 2, tell us how much of each of the building blocks we need. Note that $\hat{i}$ and $\hat{j}$ have the properties that (a) they are perpendicular (i.e., dot product is zero) (b) their magnitudes are 1, and (c) you can build any two-dimensional vector as a weighted sum of these building blocks. When a set of vectors shares these three properties, they are collectively known as an *orthonormal basis* [ortho = perpendicular, normal = magnitude of 1, basis = can be used to build every 2D vector]. Now suppose we were to consider some other vector, $\vec{y}$. As with $\vec{x}$, we would express $\vec{y}$ as a weighted sum of $\hat{i}$ and $\hat{j}$. So vectors $\vec{x}$ and $\vec{y}$ would have the building blocks $\hat{i}$ and $\hat{j}$ in common, but would differ in the weights of each of those building blocks. In fact, all 2D vectors will have the same building blocks ($\hat{i}$ and $\hat{j}$), but will differ in the weights. This means that the weights essentially describe a vector; if you know (or can solve for) the weights, then you know all there is to know about the vector. It turns out that constructing signals from weighted sums of cosines is essentially the same concept.

Our goal is to take any periodic signal $x(t)$ and represent it as a sum of cosines, or equivalently as a sum of complex exponentials. In other words we want to represent our signal as being constructed from building blocks of the form $e^{j\omega t}$ and $e^{-j\omega t}$. Although we won't go into the math to prove it, the two building blocks $e^{j\omega t}$ and $e^{-j\omega t}$ have the three properties described above for $\hat{i}$ and $\hat{j}$: they are unit magnitude, orthogonal, and they can be used to construct any periodic function $x(t)$. These building blocks must be scaled by weights that indicate how much of the building blocks we need. As we've learned in Chapter 1, those weights are the complex coefficients A and A^*, respectively.

Remember, our goal is to think about signals as being functions of frequency instead of as functions of time. This is exactly what happens when we take a signal $x(t)$ and express it as a sum of complex exponentials: the weights A and A^* tell us how much we have of a cosine of frequency ω. Recall from Chapter 1 that the magnitude of A tells us how big the cosine at frequency ω is, and the phase of A indicates the cosine's phase shift.

Section 2.2 Inner Product Integral

All we need now is some way of computing A. In other words, if you are given a periodic signal $x(t)$, we need some way of answering the question *how much $e^{j\omega t}$ (or equivalently, $\cos(\omega t)$) is in this signal?* The solution to this problem is the Inner Product integral. In general, the Inner Product integral tells you how much of a signal $y(t)$ is present in a signal $x(t)$. In our case, we want $y(t)$ to be the complex function $e^{j\omega t}$. In our specific case, the Inner Product integral is:

$$A = \frac{1}{T} \int_T x(t) \cdot e^{-j\omega t}\, dt \qquad\qquad \text{[2-1]}$$

where $x(t)$ is periodic with period $T = 2\pi/\omega$. Note that there is a more general version of the Inner Product integral that works for any signal $y(t)$, but it is beyond the scope of this book.

Example 2-1

A simple example will give us some confidence that the Inner Product integral works as advertised. Suppose $x(t) = 3\cos(t)$ and we want to use the Inner Product integral to solve for A. We can easily tell by inspection that the correct answer must be $A = 3/2$ (see Section 1.5.3).

Let's see if the Inner Product can produce this same answer. Recall that T represents the period of the signal $x(t)$, which in this case is $T = 2\pi/\omega = 2\pi$ seconds. Although in principle we can test for the presence of *any* complex exponential frequency $e^{j\omega t}$, in this case we will pick $\omega = 1$ since that is also the frequency of the signal $x(t)$.

$$A = \frac{1}{T}\int_T x(t)e^{-j\omega t}\, dt$$

$$= \frac{1}{2\pi}\int_0^{2\pi} 3\cos(t)e^{-jt}\, dt$$

$$= \frac{1}{2\pi}\frac{3}{2}\int_0^{2\pi}\left(e^{jt} + e^{-jt}\right)e^{-jt}\, dt$$

$$= \frac{3}{4\pi}\int_0^{2\pi} 1 + e^{-j2t}\, dt$$

$$= \frac{3}{4\pi}\left[t + \frac{1}{-2j}e^{-j2t}\right]_0^{2\pi}$$

$$= \frac{3}{4\pi}\left[\left(2\pi + \frac{1}{-2j}e^{-j4\pi}\right) - \left(0 + \frac{1}{-2j}\right)\right]$$

$$= \frac{3}{4\pi}\left[2\pi + \frac{1}{-2j} - 0 - \frac{1}{-2j}\right]$$

$$= \frac{3}{2} \qquad\qquad\qquad [2\text{-}2]$$

Happily, the derivation tells us what we already knew would be the right answer.

Let's now repeat the experiment for the same $x(t) = 3\cos(t)$. However, this time, instead of testing to see how much $\omega = 1$ is in $x(t)$, we'll try some other frequency, say $\omega = 2$ rads/sec. Again, we can guess the correct answer ahead of time. In this case, we'll be asking "how much $\cos(2t)$ is in the signal $x(t) = \cos(t)$?" The answer *has* to be $A = 0$: there is no element at $\omega = 2$ in a cosine whose frequency is $\omega = 1$ because a cosine, by definition, consists of energy at one and only one frequency. Let's try the calculus and see if this gets us what we expect. Don't forget that the value of T corresponds to the period of $x(t)$, which is still 2π in this case.

$$A = \frac{1}{T}\int_T x(t)e^{-j\omega t}\, dt$$

$$= \frac{1}{2\pi}\int_0^{2\pi} 3\cos(t)e^{-j2t}\, dt$$

$$= \frac{1}{2\pi}\frac{3}{2}\int_0^{2\pi}\left(e^{jt} + e^{-jt}\right)e^{-j2t}\, dt$$

$$= \frac{3}{4\pi} \int_0^{2\pi} e^{-jt} + e^{-j3t}\, dt$$

$$= \frac{3}{4\pi} \left[\frac{1}{-j} e^{-jt} + \frac{1}{-3j} e^{-j3t} \right]_0^{2\pi}$$

$$= \frac{3}{4\pi} \left[\left(\frac{1}{-j} e^{-j2\pi} + \frac{1}{-3j} e^{-j6\pi} \right) - \left(\frac{1}{-j} + \frac{1}{-3j} \right) \right]$$

$$= \frac{3}{4\pi} \left[\frac{1}{-j} + \frac{1}{-3j} - \frac{1}{-j} - \frac{1}{-3j} \right]$$

$$= 0 \hspace{6cm} [2\text{-}3]$$

which is exactly as predicted.

Section 2.3 Signal Decomposition

In the previous section, we used the function $x(t) = 3\cos(t)$ to test the ability of the Inner Product integral to compute how much of a given frequency ω is in a signal. In this section, we repeat the exercise using a square wave for $x(t)$ instead of a simple cosine. This will introduce the concept of harmonics, which is central to the study of periodic signals.

2.3.1 Square Wave

Consider the square wave function $x(t)$ shown in Figure 2-2

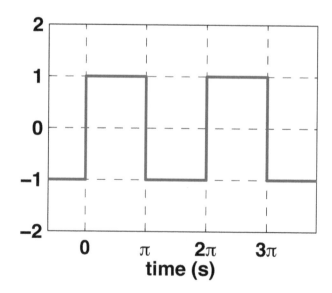

Figure 2-2: A square wave of period $T = 2\pi$ seconds.

The period of $x(t)$ is $T = 2\pi$ seconds and therefore its frequency is $\omega = 2\pi/T = 1$ rad/sec. Suppose we wish to learn how much $\cos(1t)$ exists in $x(t)$. To find out, we simply apply the Inner Product integral.

$$A = \frac{1}{T} \int_T x(t) e^{-j\omega t} \, dt$$

$$= \frac{1}{2\pi} \int_0^\pi 1 \cdot e^{-jt} \, dt + \frac{1}{2\pi} \int_\pi^{2\pi} (-1) e^{-jt} \, dt$$

$$= \frac{1}{2\pi} \frac{1}{-j} \left[e^{-jt} \right]_0^\pi - \frac{1}{2\pi} \frac{1}{-j} \left[e^{-jt} \right]_\pi^{2\pi}$$

$$= \frac{1}{-j2\pi} \left[e^{-j\pi} - e^0 \right] - \frac{1}{-j2\pi} \left[e^{-j2\pi} - e^{-j\pi} \right]$$

$$= \frac{1}{-j2\pi} [-1 - 1 - 1 - 1]$$

$$= \frac{2}{j\pi} = \frac{2}{\pi}(-j) = \frac{2}{\pi} e^{-j\frac{\pi}{2}}$$

[2-4]

What to make of this? Remember from before that we learned that $K\cos(\omega t + \phi) = Ae^{j\omega t} + A^* e^{-j\omega t}$ provided that $A = \frac{k}{2} e^{j\phi}$. We can use this property to finally write the expression for the cosine that best fits our signal $x(t)$ at $\omega = 1$ rads/sec. We see that $K = 2 \cdot |A| = 4/\pi$, and $\varphi = \angle A = -\pi/2$. Therefore, the expression we seek is:

$$x_1(t) = \frac{4}{\pi} \cos\left(t - \frac{\pi}{2} \right)$$

[2-5]

At this point, our best option is to plot Equation [2-5] to see if it makes any sense. We'll superimpose it on the original square wave $x(t)$.

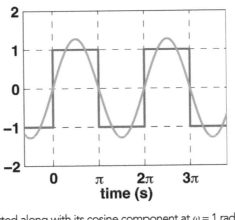

Figure 2-3: Square wave x(t) [blue] plotted along with its cosine component at $\omega = 1$ rads/sec [green]: $x_1(t) = \frac{4}{\pi} \cos\left(t - \frac{\pi}{2} \right)$.

Figure 2-3 indicates that the value of A we derived via the Inner Product integral yields a sinusoid whose amplitude and phase shift seem to match $x(t)$ reasonably well. This is good.

We now seek to repeat the derivation we've just completed, but now instead of looking for $\omega = 1$ inside of $x(t)$, we'll look to see if there is any $\omega = 2$. In general, we will test only frequencies ω that are integer multiples n of $2\pi/T$. It turns out that if we were to try applying the Inner Product integral using any frequency other than $2\pi n/T$, the resulting coefficient A would be zero. There is a lengthy (and thoroughly boring) derivation that proves this fact; for the sake of brevity, it is omitted here.

To determine the presence of energy at $\omega = 2$ rads/sec, we repeat the Inner Product integral. Note that T hasn't changed—it is still $T = 2\pi$ seconds, since it represents the period of $x(t)$.

$$A = \frac{1}{T}\int_T x(t)e^{-j\omega t}\, dt$$

$$= \frac{1}{2\pi}\int_0^\pi 1\cdot e^{-j2t}dt + \frac{1}{2\pi}\int_\pi^{2\pi}(-1)e^{-j2t}dt$$

$$= \frac{1}{2\pi}\frac{1}{-2j}\left[e^{-j2t}\right]_0^\pi - \frac{1}{2\pi}\frac{1}{-2j}\left[e^{-j2t}\right]_\pi^{2\pi}$$

$$= \frac{1}{2\pi}\frac{1}{-2j}\left[e^{-j2\pi} - 1 - e^{-j4\pi} + e^{-j2\pi}\right]$$

$$= \frac{1}{2\pi}\frac{1}{-2j}\left[1 - 1 - 1 + 1\right]$$

$$= 0 \tag{2-6}$$

So in this case, the Inner Product integral tells us that there is no energy at $\omega = 2$ rads/sec in our signal $x(t)$.

Let's try one more frequency—this time $\omega = 3$ rads/sec.

$$A = \frac{1}{T}\int_T x(t)e^{-j\omega t}\, dt$$

$$= \frac{1}{2\pi}\int_0^\pi 1\cdot e^{-j3t}dt + \frac{1}{2\pi}\int_\pi^{2\pi}(-1)e^{-j3t}dt$$

$$= \frac{1}{2\pi}\frac{1}{-3j}\left[e^{-j3t}\right]_0^\pi - \frac{1}{2\pi}\frac{1}{-3j}\left[e^{-j3t}\right]_\pi^{2\pi}$$

$$= \frac{1}{2\pi}\frac{1}{-3j}\left[e^{-j3\pi} - 1 - e^{-j6\pi} + e^{-j3\pi}\right]$$

$$= \frac{1}{2\pi}\frac{1}{-3j}\left[-1 - 1 - 1 - 1\right]$$

$$= -\frac{4}{-6j\pi}$$

$$= -\frac{2j}{3\pi}$$

$$= \frac{2}{3\pi}e^{-\frac{j\pi}{2}} \qquad [2\text{-}7]$$

Following the first example, we see that $K = 2 \cdot |A| = 4/3\pi$ and $\varphi = \angle A = -\pi/2$, and therefore

$$x_3(t) = \frac{4}{3\pi}\cos\left(3t - \frac{\pi}{2}\right) \qquad [2\text{-}8]$$

This is interesting—in the $\omega = 1$ case, the amplitude of the cosine was $4/\pi$. Now in the $\omega = 3$ case, the cosine amplitude is $4/3\pi$. We can interpret this to mean that our square wave $x(t)$ has three times more energy at $\omega = 1$ than at $\omega = 3$. This is a useful finding.

Now lets add Equation [2-8] to our plot from before; Figure 2-4 shows the result.

This graphically confirms what the numbers already told us: that there is three times as much energy at $\omega = 1$ than at $\omega = 3$ rads/sec. Also note that the $\omega = 3$ cosine has three periods for every one period of the $\omega = 1$ signal, which is exactly what we'd expect since their frequencies differ by a factor of three.

It turns out there is a more interesting way to plot the information in Figure 2-4. Instead of plotting the two cosines separately, we plot their *sum*, which is given by

$$\hat{x}(t) = \frac{4}{\pi}\cos\left(t - \frac{\pi}{2}\right) + \frac{4}{3\pi}\cos\left(3t - \frac{\pi}{2}\right) \qquad [2\text{-}9]$$

The plot of Equation [2-9] is shown in Figure 2-5.

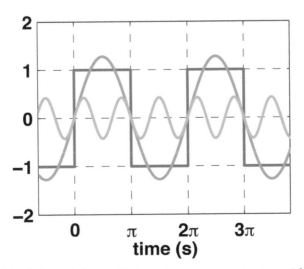

Figure 2-4: Square wave $x(t)$ [blue] plotted along with its cosine components at $\omega = 1$ [green] and 3 [cyan] rads/sec.

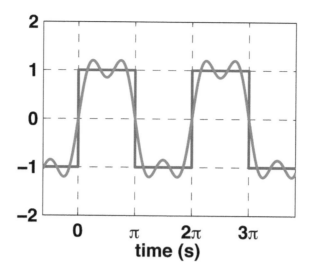

Figure 2-5: Square wave x(t) [blue] plotted along with the sum of its first two harmonics at frequencies $\omega = 1$ and 3 rads/sec [purple]. The purple trace is simply the plot of Equation [2-9].

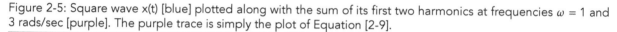

Figure 2-5 shows us that the sum of the first two cosines has started to look somewhat like the square wave itself. This is an amazing property! The only reason this has worked out so nicely is that we carefully derived the correct values of A for the cases of $\omega = 1$, 2, and 3 rads/sec. If we had used any different values for the magnitudes and phases of A, these plots wouldn't have summed together to create something that starts to look like the square wave.

To start building some intuition, try to guess how the plot in Figure 2-5 would change if we added higher frequency harmonics, say $\omega = 4$, 5, 6, and 7 rads/sec. What if we went all the way to infinity? What would the resulting sum of cosines look like, and what would the A coefficients tell us?

2.3.2 Harmonics

In the previous section, we introduced the concept of fitting only cosines to our square wave whose frequencies are integer multiples of the frequency of x(t). In the case of our sample square wave, the frequency of x(t) is $\omega = 2\pi/T = 1$ rad/sec. Therefore we are interested in fitting cosines of frequencies $\omega = 1$, 2, 3,.... . These are the *harmonics* of x(t)—the cosines of *integer multiples* of $2\pi/T$. In the previous section, we used a separate integral to solve for A for each harmonic. While effective, this was a cumbersome process, especially if we are interested in solving for the coefficient A for many higher-order harmonics.

Fortunately, there is a nice shortcut that will allow us to solve for the coefficient A for *all* of the harmonics of the square wave using just a single integral. The trick is to solve for A_n, the coefficient of the nth harmonic (where n = 1, 2, 3,...). Note that if $\omega = 2\pi/T$ is the frequency of the first harmonic, then $\omega_n = 2\pi n/T$ is the frequency of the nth harmonic. Putting this together, we can now replace the Inner Product integral of Equation [2-1] with the more general version:

$$A_n = \frac{1}{T} \int_T x(t) e^{-j\frac{2\pi n}{T}t} \, dt \qquad [2\text{-}10]$$

Equation [2-10] has been generalized so that we test *all* frequencies of cosines that are integer multiples of the frequency of $x(t)$. We can now use Equation [2-10] to calculate the magnitude and phase of every single cosine that might be relevant to the square wave $x(t)$.

$$A_n = \frac{1}{T} \int_T x(t) e^{-j\frac{2\pi n}{T} t} \, dt$$

$$= \frac{1}{2\pi} \int_0^{\pi} 1 \cdot e^{-jnt} \, dt + \frac{1}{2\pi} \int_{\pi}^{2\pi} (-1) e^{-jnt} \, dt$$

$$= \frac{1}{2\pi} \frac{1}{(-jn)} \left[e^{-jnt} \right]_0^{\pi} - \frac{1}{2\pi} \frac{1}{(-jn)} \left[e^{-jnt} \right]_{\pi}^{2\pi}$$

$$= \frac{1}{2\pi} \frac{1}{(-jn)} \left[e^{-jn\pi} - 1 - e^{-jn2\pi} + e^{-jn\pi} \right] \tag{2-11}$$

Simplifying Equation [2-11] requires noticing that the expression in brackets will take on different values depending on whether n is even or odd. When n is even, we get

$$A_n = \frac{1}{2\pi} \frac{1}{(-jn)} [1 - 1 - 1 + 1]$$

$$= 0 \tag{2-12}$$

When n is odd, we get

$$A_n = \frac{1}{2\pi} \frac{1}{(-jn)} [-1 - 1 - 1 - 1]$$

$$= -\frac{4}{2\pi} \frac{1}{(-jn)}$$

$$= -\frac{2}{\pi n} j$$

$$= \frac{2}{\pi n} e^{-\frac{j\pi}{2}} \tag{2-13}$$

Collectively, Equations [2-12] and [2-13] tell us exactly how much cosine there is at various frequencies. For example, for $n = 1, 2, \ldots, 5$, we see that

$$A_1 = \frac{2}{\pi} e^{-\frac{j\pi}{2}}$$

$$A_2 = 0$$

$$A_3 = \frac{2}{3\pi} e^{-\frac{j\pi}{2}}$$

$$A_4 = 0$$

$$A_5 = \frac{2}{5\pi} e^{-\frac{j\pi}{2}}$$

[2-14]

Recalling that the weighted sum of complex exponentials is just a fancy way of expressing cosines, we can interpret our results as

$$x_1(t) = \frac{4}{\pi} \cos\left(t - \frac{\pi}{2}\right)$$

$$x_2(t) = 0$$

$$x_3(t) = \frac{4}{3\pi} \cos\left(3t - \frac{\pi}{2}\right)$$

$$x_4(t) = 0$$

$$x_5(t) = \frac{4}{5\pi} \cos\left(5t - \frac{\pi}{2}\right)$$

[2-15]

2.3.3 Sum of Cosines

The cosines of Equation [2-15] are plotted individually in Figure 2-6(A) and are summed together in Figure 2-6(B).

As expected, we see that the sum of these first cosines starts to converge on the plot of our original square wave $x(t)$.

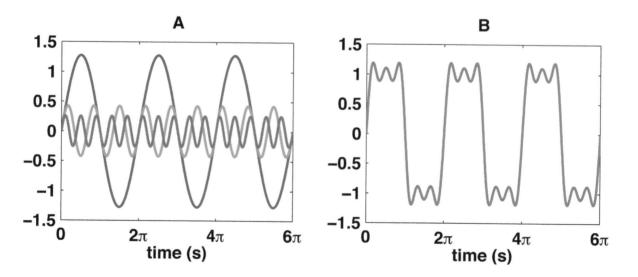

Figure 2-6: (A) The three sine waves $x_1(t)$ [blue], $x_3(t)$ [green], and $x_5(t)$ [red]. (B) The same three sine waves summed together.

The important difference between this section and the previous one is that if we want to solve for more harmonics, we don't need to re-solve the integral. Instead we just consult Equation [2-13]. For example, we can tell that the 31st harmonic is simply $x_{31}(t) = \frac{4}{31\pi} \cos\left(31t - \frac{\pi}{2}\right)$.

Now that we have a formula for A_n we can investigate what the sum of harmonics looks like as n gets larger. Figure 2-7 shows the sum of harmonics through $n = 5, 11, 25,$ and 101. Figure 2-7 shows us a couple of important things. First, we verify that as $n \to \infty$, the sum of harmonics of $x(t)$ begins to look more and more like $x(t)$ itself. In fact, what we are seeing is a visual validation of our first important theorem of this course: any signal $x(t)$ that is periodic with period T can be expressed as a sum of cosines, or equivalently a sum of complex exponentials at integer multiple (harmonic) frequencies

$$x(t) = \sum_{n=1}^{\infty} A_n e^{j\frac{2\pi n}{T}t} + A_n^* e^{-j\frac{2\pi n}{T}t} \qquad [2\text{-}16]$$

where

$$A_n = \frac{1}{T} \int_T x(t) e^{-j\frac{2\pi n}{T}t} \, dt \qquad [2\text{-}17]$$

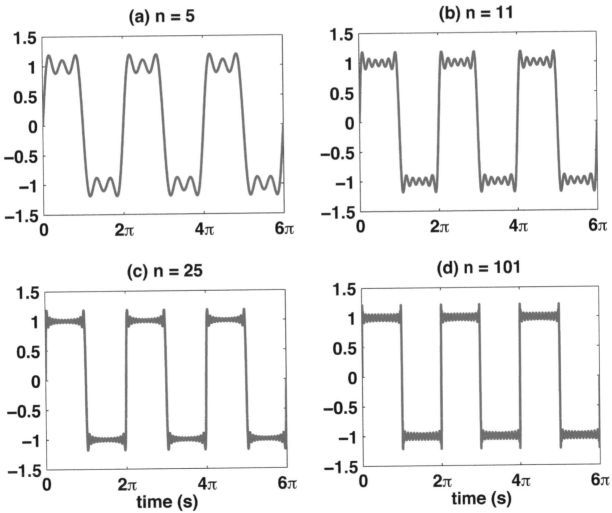

Figure 2-7: Sums of progressively higher harmonics. We see that as $n \to \infty$, the sum of harmonics converges to $x(t)$.

This is the **Fourier Series**; it is an incredibly powerful concept and it provides us with our first tool for converting signals (right now we are limited to periodic signals) from being functions of time to being functions of frequency. In other words, we are breaking our signal up into frequencies: A_n tells us how strongly each frequency is represented in $x(t)$.

The second important thing we can learn from Figure 2-7 is some intuition regarding how different frequencies contribute to a signal. Our original square wave $x(t)$ is composed of flat parts followed by instantaneous jumps (that is, it jumps from +1 to −1 and then back again). In looking at Figure 2-7 we note that we are plotting only the smallest harmonics (i.e., the lowest frequencies). The figure shows that the flat parts of the square wave are being pretty well approximated by the sum of cosines, whereas the edges of the square wave are not. This would imply that the flat parts of the square wave represent the low-frequency portions of the signal (those are the frequencies we've included in Figure 2-7), whereas the edges of the square wave must represent high-frequency portions of the signal. Since we haven't included the very highest frequencies in Figure 2-7, it stands to reason that we would expect the sum-of-cosines to match $x(t)$ the worst at the edges of the square wave; this is exactly what we observe.

2.3.4 Line Spectra Plots

A common practice is to use a graph called a line spectra plot to display the values of A_n. Note that the values of A_n are typically complex, which means that we will need two graphs to properly display all the information: a magnitude plot and a phase plot. For the sake of clarity, we summarize the process in Table 2-1 using the results from the square wave example from Section 2.3.1.

Table 2-1: Coefficients for the Fourier series example in Section 2.3.1

ω (rad/sec)	A_n	$|A_n|$	$\angle A_n$
0	0	0	0
1	$\frac{2}{\pi}e^{-j\pi/2}$	$\frac{2}{\pi}$	$-\pi/2$
2	0	0	0
3	$\frac{2}{3\pi}e^{-j\pi/2}$	$\frac{2}{3\pi}$	$-\pi/2$
4	0	0	0
5	$\frac{2}{5\pi}e^{-j\pi/2}$	$\frac{2}{5\pi}$	$-\pi/2$
6	0	0	0
7	$\frac{2}{7\pi}e^{-j\pi/2}$	$\frac{2}{7\pi}$	$-\pi/2$

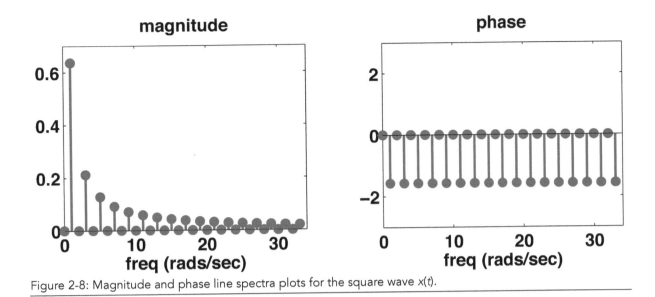

Figure 2-8: Magnitude and phase line spectra plots for the square wave $x(t)$.

From Table 2-1, it is trivial to create the magnitude and phase line spectra plots, as shown in Figure 2-8. From the magnitude plot, we can see that there is progressively less and less energy in the square wave as the frequency increases. That plot shows clearly why there is a big improvement in the shape of the signal when we add the first few harmonics (because those first harmonics have a lot of energy), but relatively little improvement as we add each additional harmonic. The phase plot shows what we already knew: that the phase angle of every non-zero cosine is $-\pi/2$ radians.

2.3.5 Magnitude and Phase

Next let's work a simple example to build some intuition about the differences in information embedded in the magnitude and phase plots. Let's find the Fourier Series for the square wave $y(t)$ shown in Figure 2-9. This square wave has the same amplitude and frequency as $x(t)$. It differs only in terms of a right shift relative to $x(t)$ of $\pi/3$ radians.

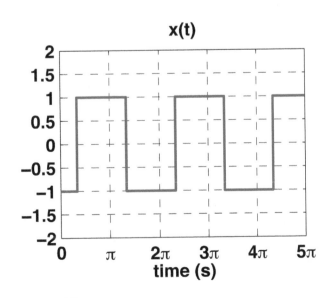

Figure 2-9: Square wave $y(t) = x(t - \pi/3)$.

We solve for the Fourier Series as before.

$$A_n = \frac{1}{T}\int_T y(t)e^{-j\frac{2\pi n}{T}t}\,dt$$

$$= \frac{1}{2\pi}\int_{\pi/3}^{4\pi/3} 1\cdot e^{-jnt}\,dt + \frac{1}{2\pi}\int_{4\pi/3}^{7\pi/3}(-1)e^{-jnt}\,dt$$

$$= \frac{1}{2\pi}\frac{-1}{jn}\left(\left[e^{-jnt}\right]_{\pi/3}^{4\pi/3} - \left[e^{-jnt}\right]_{4\pi/3}^{7\pi/3}\right)$$

$$= \frac{j}{2\pi n}\left(e^{-j4\pi n/3} - e^{-j\pi n/3} - e^{-j7\pi n/3} + e^{-j4\pi n/3}\right) \qquad \text{[2-18]}$$

There are several complex number identities that could be applied to simplify Equation [2-18]. Regardless of which is used, we are left with

$$A_n = \begin{cases} \dfrac{2}{\pi n}e^{-j5\pi/6} & \text{if } n=1,7,13,\ldots \\[2mm] \dfrac{2}{\pi n}e^{j\pi/2} & \text{if } n=3,9,15,\ldots \\[2mm] \dfrac{2}{\pi n}e^{-j\pi/6} & \text{if } n=5,11,17,\ldots \\[2mm] 0 & \text{if } n \text{ is even} \end{cases} \qquad \text{[2-19]}$$

Figure 2-10 shows the line spectra plots.

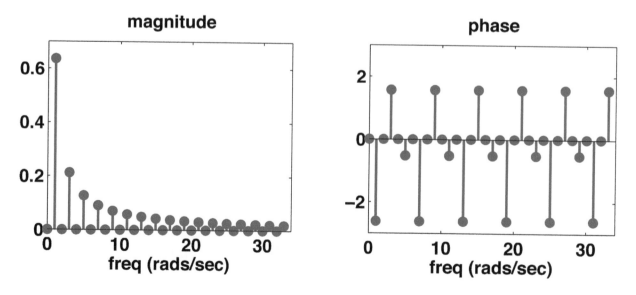

Figure 2-10: Magnitude and phase line spectra plots for the square wave $y(t)$.

Comparing Figure 2-8 and Figure 2-10, we notice something very interesting. The functions x(t) and y(t) have *exactly* the same magnitude spectra. They differ only in their phase spectra. Understanding why this is so will add a lot to your intuitive understanding of the spectra. In comparing x(t) and y(t), it seems reasonable that both would have the same amount of energy at $\omega = 1$ rads/sec since both have the same amplitude, shape, and frequency. If we are trying to express x(t) and y(t) as a sum of cosines, we would need precisely the same amount of signal at $\omega = 1$ rads/sec. The only difference would be that for y(t), we would need to add some phase angle to the cosine in order to shift it to the right. This is exactly what is shown in Figure 2-8 and Figure 2-10. The magnitude tells us *how much* of a particular cosine we have, whereas the phase tells us what the *shift* is for each particular frequency.

2.3.6 Stair-Step Function

Let's practice deriving the Fourier Series by hand. This is good practice and it will develop some valuable intuition along the way. Remember, any periodic signal can be expressed as a sum of cosines whose frequencies are integer multiples of $2\pi/T$.

We start with the stair-step function x(t) shown in Figure 2-11. Our goal is to represent this function as a sum of cosines. We begin by determining the period T, which is apparent by inspection: $T = 3$ seconds. Next, we apply the Inner Product integral and tackle the calculus.

$$A_n = \frac{1}{T}\int_T x(t)e^{-j\frac{2\pi n}{T}t}\,dt$$

$$= \frac{1}{3}\int_0^1 1\cdot e^{-j\frac{2\pi n}{3}t}\,dt + \frac{1}{3}\int_1^2 2\cdot e^{-j\frac{2\pi n}{3}t}\,dt + \frac{1}{3}\int_2^3 0\cdot e^{-j\frac{2\pi n}{3}t}\,dt$$

$$= -\frac{1}{j2\pi n}\left[e^{-j\frac{2\pi n}{3}t}\right]_0^1 + -\frac{1}{j\pi n}\left[e^{-j\frac{2\pi n}{3}t}\right]_1^2$$

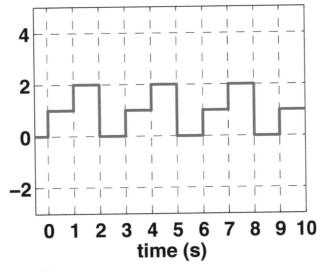

Figure 2-11: Stair-Step function x(t).

$$= -\frac{1}{j2\pi n}\left(e^{-\frac{j2\pi n}{3}} - 1\right) - \frac{1}{j\pi n}\left(e^{-\frac{j4\pi n}{3}} - e^{-\frac{j2\pi n}{3}}\right)$$

$$= \frac{1}{2\pi n}\left(e^{-\frac{j2\pi n}{3}} - 2e^{-\frac{j4\pi n}{3}} + 1\right)e^{-\frac{j\pi}{2}} \qquad [2\text{-}20]$$

With some effort at simplifying the complex terms we wind up with

$$A_n = \begin{cases} \dfrac{3}{2\pi n}e^{-\frac{j5\pi}{6}} & \text{if } n = 1, 4, 7,\dots \\[2mm] \dfrac{3}{2\pi n}e^{-\frac{j\pi}{6}} & \text{if } n = 2, 5, 8,\dots \\[2mm] 0 & \text{if } n = 3, 6, 9,\dots \end{cases} \qquad [2\text{-}21]$$

The resulting plots are shown in Figure 2-12.

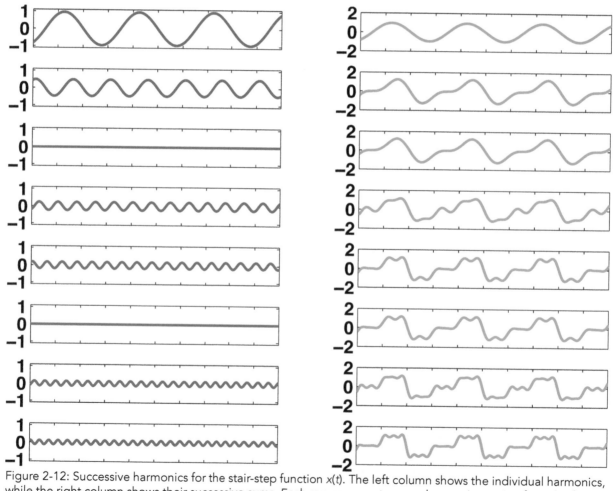

Figure 2-12: Successive harmonics for the stair-step function $x(t)$. The left column shows the individual harmonics, while the right column shows their successive sums. Each row represents a new harmonic starting from the first to the eighth. The time axes show 0-10 seconds.

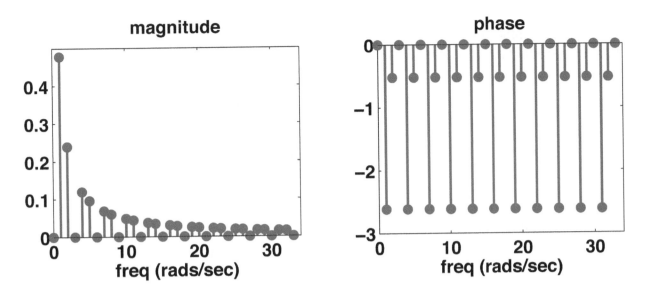

Figure 2-13: Line spectra for signal x(t).

Figure 2-12 clearly shows that even with just eight harmonics (two of which are zero, no less), we are able to represent the signal x(t) reasonably well as a sum of cosines. Figure 2-13 shows the corresponding magnitude and phase line spectra.

As in the previous section, we see the most of the energy is added at low frequencies, with progressively less energy from higher ones. Figure 2-13 is also consistent with Figure 2-12 in that the first few harmonics (the lower frequencies) give the general shape of the flat portions of x(t). The higher harmonics improve the shape of the signal around the high-frequency edges of x(t).

Section 2.4 Properties of the Fourier Transform

There are a number of important properties of the Fourier Series that we can exploit if we are careful. Here we will summarize those properties and then give an example of how they can be combined. In all cases, we assume that the Fourier Series for some generic signal x(t) is given by

$$x(t) \Leftrightarrow A_n = \frac{1}{T} \int_0^T x(t) e^{-j\frac{2\pi n}{T}t} \, dt \qquad [2\text{-}22]$$

2.4.1 DC Offset Property

A close inspection of Figure 2-12 indicates that our solution does not precisely reproduce the signal x(t) of Figure 2-11. In fact, our solution is shifted down by one unit relative to x(t). The reason for this leads us to an important observation about the Fourier Series. Clearly, we need to add the constant value of 1 to our Fourier approximation of x(t). It turns out that you can think of constant functions as being cosines with frequency zero. In other words, if $\omega = 0$, then $K\cos(\omega t) = K\cos(0) = K$. This means that we can solve for the vertical offset by solving for $A_{n=0}$, which is the coefficient corresponding to $\omega = \frac{2\pi}{T} \cdot 0 = 0$ rads/sec. We can solve for A_0 by solving the Inner Product integral as shown in Equation [2-23].

$$A_0 = \frac{1}{3} \int_0^3 x(t) e^{-j0t} \, dt$$

$$= \frac{1}{3} \int_0^3 x(t) \, dt$$

$$= 1 \qquad\qquad\qquad [2\text{-}23]$$

Happily, we get the answer we expect. Note that the value of A_0 always equals the average value of signal $x(t)$ over one period. Also, since zero-frequency signals are often associated with "Direct Current" voltage sources, we refer to them as "DC offsets."

Another way to think about this example is that adding a DC offset to your signal does not affect any of the harmonics: it doesn't affect the actual frequencies themselves or the values of A_n. Adding a DC offset changes only A_0. This should make sense intuitively. Moving a signal up or down does not affect how much signal there is at the various frequencies. Therefore it is reasonable to expect that the A_n values won't change.

2.4.2 Scaling Property

Suppose $x(t)$ is scaled by a constant c. The Fourier Series becomes

$$A_n^{new} = \frac{1}{T} \int_0^T cx(t) e^{-j\frac{2\pi n}{T}t} \, dt$$

$$= c \frac{1}{T} \int_0^T x(t) e^{-j\frac{2\pi n}{T}t} \, dt$$

$$= cA_n \qquad\qquad\qquad [2\text{-}24]$$

In other words, if we scale a function $x(t)$ by a constant, then we also scale the Fourier Series by the same constant. This makes sense since the phase of the Fourier Series is unchanged but the magnitude is scaled by c. Scaling a function $x(t)$ should not change any of the phase relationships between the various frequencies; it should change only *how much* of each of those frequencies we have.

2.4.3 Summation Property

Suppose we have a signal $x(t)$ that can be expressed as a sum of two other signals $x_a(t)$ and $x_b(t)$. If we know the Fourier Series for $x_a(t)$ and $x_b(t)$, what can we infer about the Fourier Series of $x(t)$? In other words, assume we know $x(t) = x_a(t) + x_b(t)$, and that $x_a(t) \Leftrightarrow A_n^a$, $x_b(t) \Leftrightarrow A_n^b$ and that all three x signals are periodic with period T. The following analysis describes A_n in terms of A_n^a and A_n^b.

$$A_n = \frac{1}{T} \int_0^T x(t) e^{-j\frac{2\pi n}{T}t} \, dt$$

$$= \frac{1}{T} \int_0^T \left(x_a(t) + x_b(t)\right) e^{-j\frac{2\pi n}{T}t} \, dt$$

$$= \frac{1}{T} \int_0^T x_a(t) e^{-j\frac{2\pi n}{T}t} \, dt + \frac{1}{T} \int_0^T x_b(t) e^{-j\frac{2\pi n}{T}t} \, dt$$

$$= A_n^a + A_n^b \qquad \text{[2-25]}$$

So the Fourier Series of a sum of two functions is the sum of their respective Fourier Series.

2.4.4 Time-Shift Property

Suppose we shift $x(t)$ to the right by τ seconds. In other words, suppose we want the Fourier Series of $x(t-\tau)$. The following derivation shows how the Fourier Transform of $x(t)$ is affected by the shift.

$$A_n^{new} = \frac{1}{T} \int_0^T x(t-\tau) e^{-j\frac{2\pi n}{T}t} \, dt$$

$$= \frac{1}{T} \int_T x(u) e^{-j\frac{2\pi n}{T}(u+\tau)} \, du$$

$$= \frac{1}{T} \int_T x(u) e^{-j\frac{2\pi n}{T}u} e^{-j\frac{2\pi n}{T}\tau} \, du$$

$$= e^{-\frac{j2\pi n}{T}\tau} \frac{1}{T} \int_T x(u) e^{-j\frac{2\pi n}{T}u} \, du$$

$$= e^{-j\frac{2\pi n}{T}\tau} A_n \qquad \text{[2-26]}$$

Therefore if we shift a signal in time by τ seconds, the magnitude of the Fourier Series remains unchanged. Instead we simply subtract $2\pi n\tau/T$ radians from the phase. This should make sense intuitively, since shifting a signal in time does not affect the amount (i.e., magnitude) of each frequency we have.

Example 2-2

Recall the stair-step signal from before (Figure 2-11). By inspection, we can see that this signal can be broken up into the sum of two separate signals: $x(t) = x_a(t) + x_b(t)$ (see Figure 2-14). Use the properties presented thus far to produce an alternate derivation of the Fourier Series of the stair-step signal.

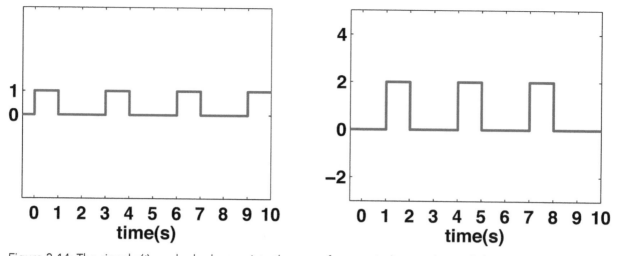

Figure 2-14: The signal $x(t)$ can be broken up into the sum of two periodic signals $x_a(t)$ [left] and $x_b(t)$ [right].

Solution: We start by solving for the Fourier Series of $x_a(t)$. Note that $T = 3$.

$$A_n = \frac{1}{3}\int_0^3 x(t)e^{-j\frac{2\pi n}{3}t}\, dt$$

$$= \frac{1}{3}\int_0^1 e^{-j\frac{2\pi n}{3}t}\, dt$$

$$= \frac{1}{3}\frac{3}{3 - j2\pi n}\left[e^{-j\frac{2\pi n}{3}t}\right]_0^1$$

$$= \frac{1}{3}\frac{3}{3 - j2\pi n}\left[e^{-j\frac{2\pi n}{3}} - 1\right]$$

$$= \frac{1}{3}\frac{3}{3 - j2\pi n}\, e^{-j\frac{\pi n}{3}}\left[e^{-j\frac{\pi n}{3}} - e^{j\frac{\pi n}{3}}\right]$$

$$= \frac{1}{\pi n}\sin\left(\frac{\pi n}{3}\right)e^{-j\frac{\pi n}{3}} \qquad\qquad [2\text{-}27]$$

Next we need to solve for the Fourier Series of $x_b(t)$. Fortunately, there is a relationship between $x_a(t)$ and $x_b(t)$. Specifically, $x_b(t) = 2x_a(t-1)$. According to the rules we have developed so far, we should be able to infer the formula for A_n^b by applying the rules for scaling and time-shifting.

$$A_n^b = 2A_n^a\, e^{-j\frac{2\pi n}{3}\cdot 1}$$

$$= \frac{2}{\pi n}\sin\left(\frac{\pi n}{3}\right)e^{-j\frac{\pi n}{3}}e^{-j\frac{2\pi n}{3}}$$

$$= \frac{2}{\pi n}\sin\left(\frac{\pi n}{3}\right)e^{-j\pi n} \qquad\qquad [2\text{-}28]$$

We complete our analysis by summing A_n^a and A_n^b to get the Fourier Series for $x(t)$.

$$A_n = A_n^a + A_n^b$$

$$= \frac{1}{\pi n}\sin\left(\frac{\pi n}{3}\right)e^{-\frac{j\pi n}{3}} + \frac{2}{\pi n}\sin\left(\frac{\pi n}{3}\right)e^{-j\pi n}$$ [2-29]

Fortunately, the expression simplifies quite nicely for different values of n.

$$A_n = \begin{cases} \dfrac{3}{2\pi n}e^{-j\frac{5\pi}{6}} & n = 1, 4, 7,\ldots \\[2ex] \dfrac{3}{2\pi n}e^{-j\frac{\pi}{6}} & n = 2, 5, 8,\ldots \\[2ex] 0 & n = 3, 6, 9,\ldots \end{cases}$$ [2-30]

which is exactly the same result we achieved in Section 2.3.6. Whether it was easier to solve the problem this way or using the approach of Section 2.3.6 is immaterial. The important thing is to realize intuitively that functions can be decomposed into sub-functions that can be interpreted more easily and then combined back together.

2.4.5 Odd and Even Functions

Consider a simple cosine function $x(t) = \cos(\omega_c t)$. This cosine has frequency ω_c radians per second and therefore a period of $2\pi/\omega_c$ seconds. We can deduce its Fourier Series expansion by simply recalling the complex exponential definition of cosine:

$$x(t) = \frac{1}{2}e^{j\omega_c t} + \frac{1}{2}e^{-j\omega_c t}$$ [2-31]

We remember that the A_n values are simply the coefficients of the $e^{j\omega t}$ terms, and therefore we conclude that in this case, there is only one value of A (corresponding to $n = 1$) and its value is 1/2:

$$A_n = \begin{cases} 1/2 & n = 1 \\ 0 & \text{all other } n \end{cases}$$ [2-32]

The plot of the Fourier Series follows easily enough.

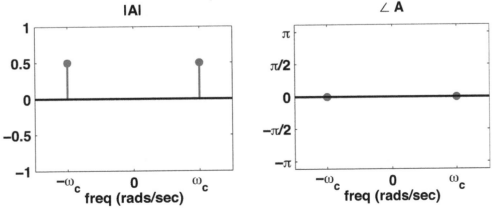

Figure 2-15: Graphical representation of the Fourier Series of $x(t) = \cos(\omega_c t)$.

Now let's say we want to calculate the Fourier Series for $x(t) = \sin(\omega_c t)$. You might recall from before that $\sin(\omega_c t) = \cos\left(\omega_c t - \frac{\pi}{2}\right) = \cos\left(\omega_c\left(t - \pi/2\omega_c\right)\right)$. Therefore, our sine wave is equivalent to a cosine that has been shifted right by $\tau = \pi/2\omega_c$ seconds. We learned in Section 2.4.4 that shifting a signal to the right in time by τ seconds is equivalent to multiplying A_n by $e^{-j\omega\tau}$. Therefore, the Fourier representation of our sine wave should be $\frac{1}{2}e^{-j\omega_c\pi/2\omega_c} = \frac{1}{2}e^{-j\pi/2}$. Or more completely:

$$A_n = \begin{cases} (1/2)e^{-j\pi/2} & n = 1 \\ 0 & all\ other\ n \end{cases}$$

[2-33]

Figure 2-16 shows this graphically. As expected, when comparing the A_n for sine and cosine, we see they have the same magnitude but different phases. This is because time shifts don't affect the amount of each frequency, just the relative location in time.

This example leads to an interesting observation. In the first case, $x(t) = \cos(\omega_c t)$ is an even function. Recall that even functions satisfy the criterion that $f(-x) = f(x)$. Equivalently, even functions are symmetric about the y-axis. The Fourier Series for the cosine was $A = 1/2$, which is purely real (no imaginary part). Now compare with the second example. In that case, $x(t) = \sin(\omega_c t)$ is an odd function. Odd functions satisfy $f(-x) = -f(x)$ and are symmetric about the line $y = x$. In that case, we found that $A = (1/2)e^{-j\pi/2} = -j/2$, which is purely imaginary (no even part). This is no coincidence. In general, any even function will have a Fourier Series that is purely real and any odd function will have a purely imaginary Fourier Series. Functions that are neither even nor odd will have complex Fourier Series, that is, containing both real and imaginary portions.

In fact, we've already seen an example of this. Recall that we calculated the Fourier Series of the odd function $x(t)$ shown in Figure 2-2. The result, shown in Equation [2-14], shows that all non-zero A_n coefficients are purely odd (because $e^{-\frac{j\pi}{2}} = -j$).

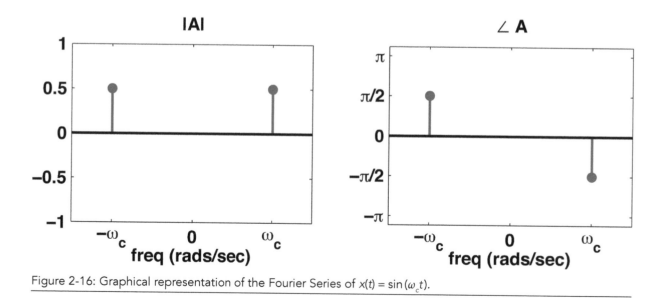

Figure 2-16: Graphical representation of the Fourier Series of $x(t) = \sin(\omega_c t)$.

As an experiment, if we were to shift the square wave of Figure 2-2 to the left by $\pi/2$ seconds, it would become an even function. This is the same as applying a right shift of $\tau = -\pi/2$ seconds. Therefore the effect on the Fourier Series is to multiply it by $e^{-j\omega\tau} = e^{-j(2\pi n/T)\tau}$. Therefore

$$A_n = \frac{2}{\pi n} e^{-j\frac{\pi}{2}} \cdot e^{-j(2\pi n/2\pi)(-\pi/2)}$$

$$= \frac{2}{\pi n} e^{-j\pi/2} \cdot e^{j\pi n/2}$$

$$= \begin{cases} \dfrac{2}{\pi n} & n = 1, 5, 9, \ldots \\[2ex] -\dfrac{2}{\pi n} & n = 3, 7, 11, \ldots \\[2ex] 0 & n \text{ is even} \end{cases} \qquad [2\text{-}34]$$

As expected, our newly created even function has a purely real Fourier Series.

2.4.6 Time-Scaling Property

Suppose we take the square wave of Figure 2-2 and scale it with respect to time so that it becomes $y(t)$ as shown in Figure 2-17.

This is exactly the same square wave as in Figure 2-2 except it is compressed in time. Mathematically, we say $y(t) = x(6t)$. Note that the period of the new signal is one sixth of the period of $x(t)$: $T_y = 2\pi/6 = T_x/6$. In order to solve for the Fourier Series of $y(t)$, we derive the time-scaling property. For the sake of generality, let $y(t) = x(at)$ where a is some constant.

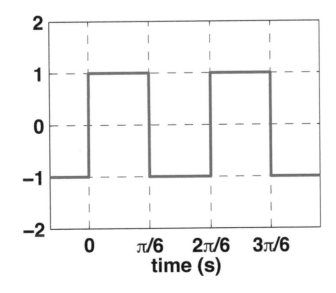

Figure 2-17: A square wave $y(t)$ with period $T = 2\pi/6$ seconds.

$$A_n^y = \frac{1}{T_y} \int_0^{T_y} y(t) e^{-j\frac{2\pi n}{T_y}t} \, dt$$

$$= \frac{a}{T_x} \int_0^{\frac{T_x}{a}} x(at) e^{-j\frac{2\pi na}{T_x}t} \, dt$$

$$\text{let } u = at \text{ and } du = a \, dt$$

$$= \frac{a}{T_x} \frac{1}{a} \int_0^{T_x} x(u) e^{-j\frac{2\pi n}{T_x}u} \, du$$

$$= A_n^x \tag{2-35}$$

This is amazing! We have discovered that time scaling the signal does not affect the Fourier Series. In other words, signals $x(t)$ and $y(t)$ have exactly the same amount of energy in each of their respective harmonics. The only difference is the frequency of the nth harmonic. The table of A_n for $y(t)$ will be exactly as it was for $x(t)$ (see Equation [2-14]). The first five cosines comprising $y(t)$ are given by

$$x_1(t) = \frac{4}{\pi} \cos\left(6t - \frac{\pi}{2}\right)$$

$$x_2(t) = 0$$

$$x_3(t) = \frac{4}{3\pi} \cos\left(18t - \frac{\pi}{2}\right)$$

$$x_4(t) = 0$$

$$x_5(t) = \frac{4}{5\pi} \cos\left(30t - \frac{\pi}{2}\right) \tag{2-36}$$

In comparison to Equation [2-15], we see that although the frequency of the nth harmonic has increased by a factor of $a = 6$, the actual coefficient values A_n have not changed. Upon reflection, this should make some intuitive sense: time scaling a signal does not alter the shape of the signal, just its frequency. Therefore we should expect that even though its harmonic frequencies will change, its Fourier coefficients (which define the signal shape) will stay the same.

Section 2.5 Summary

This chapter introduced the concept of the Fourier Series. The Fourier Series is an expression in which a periodic signal can be written as a weighted sum of cosines (or, equivalently, complex exponentials) at integer multiple (harmonic) frequencies. The tool for decomposing a periodic signal into its Fourier coefficients is the Inner Product integral. In general, for any periodic signal $x(t)$ with period T, we can say:

$$x(t) = A_0 + \sum_{n=1}^{\infty} A_n e^{j\frac{2\pi n}{T}t} + A_n^* e^{-j\frac{2\pi n}{T}t}$$

where

$$A_n = \frac{1}{T} \int_T x(t) e^{-j\frac{2\pi n}{T}t} \, dt$$

and

$$A_0 = \frac{1}{T} \int_T x(t) \, dt$$

Alternately, the sum can be expressed with cosines as:

$$x(t) = A_0 + \sum_{n=1}^{\infty} 2|A_n| \cos\left(\frac{2\pi n}{T}t + \angle A_n\right)$$

Finally, a series of properties are helpful when manipulating periodic signals.

Table 2-2: Summary of Fourier series properties

Scaling	$cx(t) \Rightarrow cA_n$
Summation	$x_a(t) + x_b(t) \Rightarrow A_n^a + A_n^b$
Time Shift	$x(t-\tau) \Rightarrow e^{\frac{j2\pi n}{T}\tau} A_n$
Even Functions	A_n is purely real
Odd Functions	A_n is purely imaginary
Time Scaling	$x(at) \Rightarrow A_n$

Chapter 3

Fourier Transform

Section 3.1 Deriving Fourier Transform from Fourier Series

S o far, we have limited our conversation to periodic signals. We have learned that any signal $x(t)$ that is periodic with period T seconds can be expressed as a sum of cosines whose frequencies are integer multiples of $2\pi/T$ radians per second. These are called the harmonic frequencies. Therefore a signal with period $T = 2\pi$ seconds has energy at frequencies of $\omega = 1, 2, 3,\ldots$ radians/sec. However, very importantly, there is zero energy at all other frequencies. For example, there is no energy at $\omega = 1.2$ rads/sec or, say, 100.87 rads/sec. For periodic signals, there is only energy at the harmonic frequencies of $2\pi n/T$.

Let's examine how to extend our understanding of Fourier Series to *aperiodic* signals. Aperiodic signals are ones that don't repeat, and therefore have no period, T. Start by considering the function in Figure 3-1.

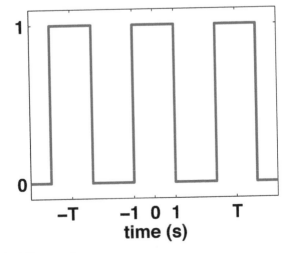

Figure 3-1: Signal $x(t)$ with period T seconds.

For this function, we're going to be clever and let the period of the function be a variable. The plan is to compute A_n as a function of the period T and then see how A_n varies with T. If we make T get bigger, that will mean the spacing between the square pulses will get larger. In the limit as $T \to \infty$, the signal $x(t)$ will no longer be periodic: instead it will be a single pulse on the range $-1 \le t < 1$.

First we solve for A_n as a function of T.

$$A_n = \frac{1}{T}\int_{-\frac{T}{2}}^{\frac{T}{2}} x(t)\, e^{-j\frac{2\pi n}{T}t}\, dt$$

$$= \frac{1}{T}\int_{-1}^{1} 1 \cdot e^{-j\frac{2\pi n}{T}t}\, dt$$

$$= \frac{1}{T}\frac{T}{-j2\pi n}\left[e^{-j\frac{2\pi n}{T}t} \right]_{-1}^{1}$$

$$= \frac{1}{-j2\pi n}\left[e^{-j\frac{2\pi n}{T}} - e^{j\frac{2\pi n}{T}} \right]$$

$$= \frac{1}{-j2\pi n}(-2j)\sin\left(\frac{2\pi n}{T}\right)$$

$$= \frac{1}{\pi n}\sin\left(\frac{2\pi n}{T}\right) \qquad\qquad [3\text{-}1]$$

Note that because $x(t)$ is a purely even function, its Fourier Series A_n is purely real. We know that the harmonics occur at frequencies that are integer multiples of $2\pi/T$. Figure 3-2 plots the harmonics of $x(t)$ assuming that $T = 4$ seconds.

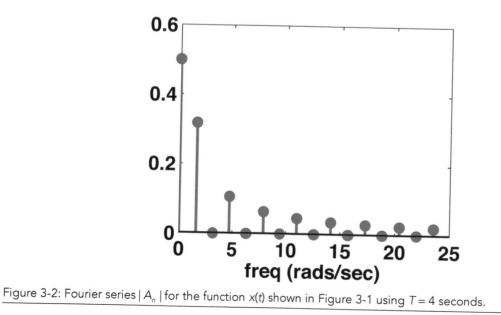

Figure 3-2: Fourier series $|A_n|$ for the function $x(t)$ shown in Figure 3-1 using $T = 4$ seconds.

Note that the harmonics are at integer multiples of $2\pi/4 \approx 1.6$ rads/sec, which explains why the first three harmonics can be seen at 1.6, 3.2, and 4.8 rads/sec.

Figure 3-3 shows the Fourier Series A_n for increasingly larger values of T. Figure 3-3 shows an interesting trend. As T gets larger, the spacing between the harmonic frequencies, $2\pi/T$, gets smaller. As the spacing between those harmonics gets smaller, it also seems that the shape that is outlined by the points is converging onto a specific pattern. We can reasonably extrapolate therefore, that as $T \to \infty$, the harmonics become infinitely close together. This means that for aperiodic signals, there is energy at every single frequency, not just at select harmonics. In other words, whereas before frequency was a discrete variable, now it is continuous. Instead of considering just the nth harmonic A_n, we are considering the energy of the signal is a continuous function of frequency, or $F(j\omega)$.

This is the essence of the *Fourier Transform*: an aperiodic signal can be expressed as a sum of cosines of *all* frequencies at once. It is an important logical jump to acknowledge that *aperiodic signals have energies over a continuous range of frequencies*.

The formal definition of the Fourier Transform is as follows. Assuming that $x(t)$ is aperiodic,

$$F(j\omega) = \int_{-\infty}^{\infty} x(t)e^{-j\omega t}dt \qquad [3\text{-}2]$$

where $F(j\omega)$ is a complex function that tells us how much "stuff" there is of $x(t)$ at each frequency.

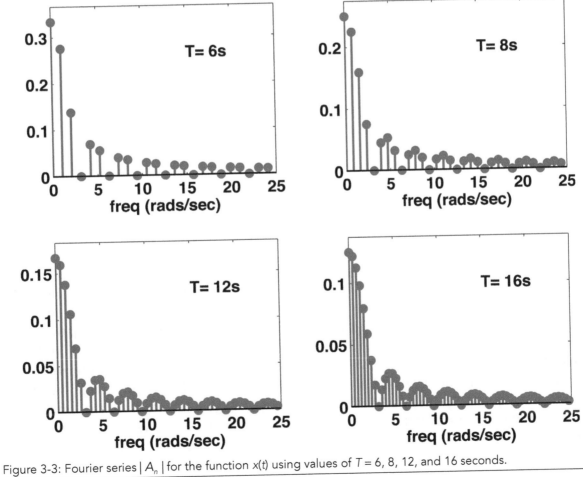

Figure 3-3: Fourier series $|A_n|$ for the function $x(t)$ using values of $T = 6, 8, 12,$ and 16 seconds.

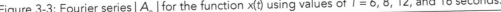

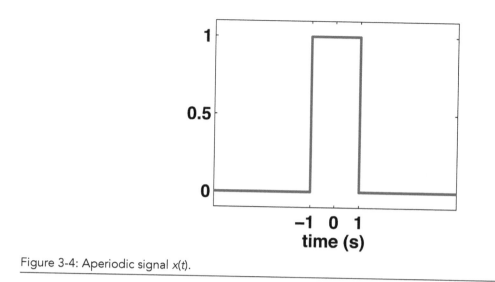

Figure 3-4: Aperiodic signal x(t).

Let's try a straightforward example. Consider the function x(t) shown in Figure 3-4. To solve for the frequency content of this signal, we apply the definition of the Fourier Transform.

$$F(j\omega) = \int_{-\infty}^{\infty} x(t)e^{-j\omega t}dt$$

$$= \int_{-1}^{1} 1 \cdot e^{-j\omega t}dt$$

$$= \frac{1}{-j\omega}\left[e^{-j\omega} - e^{j\omega}\right]$$

$$= \frac{-2j}{-j\omega}\sin(\omega)$$

$$= \frac{2\sin(\omega)}{\omega} \qquad\qquad [3\text{-}3]$$

The plot of $|F(j\omega)|$ is shown in Figure 3-5.

There are three important things to note. First, in comparing Figure 3-3 and Figure 3-5, we see that the Fourier Transform in Figure 3-5 is exactly the shape that the Fourier Series in Figure 3-3 was converging to as $T \to \infty$. This is not a coincidence! Secondly, we note that, although we've plotted only the function $F(j\omega)$ for positive values of $j\omega$, there are also points for negative values of $j\omega$. However you'll recall from our conversation of Fourier Series that the magnitude plot is always symmetric across the y-axis (i.e., even) and therefore it would be redundant to plot those points. Plotting them would just take up space without telling us anything new. Finally, we note that Figure 3-5 plots just the magnitude of $F(j\omega)$. In general, when we deal with Fourier Transforms, we will have to plot both the magnitude and the phase. However, for now, we show just the magnitude plot for simplicity in making our point about the relationship between Fourier Transforms and Fourier Series.

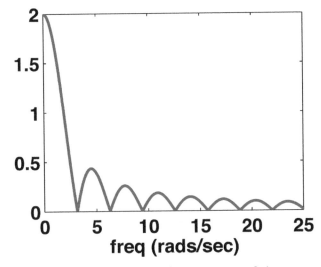

Figure 3-5: Fourier Transform $|F(j\omega)|$ of the function $x(t)$ shown in Figure 3-4.

Section 3.2 Fourier Transform of a Square Pulse

The Fourier Transform is a formula that takes a signal in time, $x(t)$, and coverts it to a signal in frequency, $F(j\omega)$. In other words, it tells us which frequencies have energy and which don't.

As much fun as it might be to spend an entire semester doing brutal integrals to solve for various Fourier Transforms, it would be a waste of time and it wouldn't yield any useful insight. Instead we are going to focus on understanding how the Fourier Transforms of different signals are related. The idea is that, if you know the Fourier Transform of one signal, you should be able to infer most of the relevant information about the Fourier Transform of some related signal.

To illustrate this point, let's start by solving for the Fourier Transform of some generic square pulse, as shown in Figure 3-6. Note that variable a is a parameter that can be varied to make

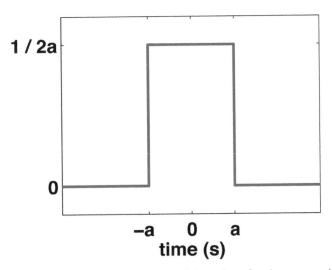

Figure 3-6: Generic square pulse $x(t)$. Note that, regardless of the value of a, the area under the pulse is always one.

the pulse wider (and shorter) or thinner (and taller). However the area of the square pulse will always be $\left(2a \times \frac{1}{2a}\right) = 1$.

The following steps solve for the Fourier Transform of $x(t)$.

$$F(j\omega) = \int_{-\infty}^{\infty} x(t) e^{-j\omega t} dt$$

$$= \int_{-a}^{a} \frac{1}{2a} e^{-j\omega t} dt$$

$$= \frac{1}{2a} \cdot \frac{1}{-j\omega} \left[e^{-j\omega t} \right]_{-a}^{a}$$

$$= \frac{1}{2a} \cdot \frac{1}{-j\omega} \left[e^{-j\omega a} - e^{j\omega a} \right]$$

$$= \frac{1}{2a} \cdot \frac{1}{-j\omega} (-2j) \sin(a\omega)$$

$$= \frac{\sin(a\omega)}{a\omega} \qquad\qquad [3\text{-}4]$$

Right off the bat, we notice something interesting about Equation [3-4]. We note that the square pulse shown in Figure 3-6 is a purely even function. We therefore expect the Fourier Transform to be purely real. We are pleased to observe that Equation [3-4] is, in fact, purely real.

Equation [3-4] is a version of the well-known *sinc* function, which is defined as

$$\text{sinc}(\omega) = \frac{\sin(\pi\omega)}{\pi\omega} \qquad\qquad [3\text{-}5]$$

By comparing Equations [3-4] and [3-5] we can rewrite Equation [3-4] as

$$F(j\omega) = \text{sinc}(a\omega / \pi) \qquad\qquad [3\text{-}6]$$

This is a handy thing to know because Matlab has a built-in sinc function that can be used to easily plot Fourier Transforms such as these.

Let's now plot Equation [3-4] (or equivalently Equation [3-6]) and see what the frequency content of our signal $x(t)$ looks like. In general, when plotting a Fourier Transform, we are obliged to produce *two* plots, one for magnitude and one for phase, because the Fourier Transform is a complex function. The following Matlab code produces the plots we seek. For our case here, we let $a = 1$.

```
w = linspace(-5,5,1e5);
F = sinc(w/pi);
subplot(1,2,1); plot(w,abs(F));
subplot(1,2,2); plot(w,angle(F));
```

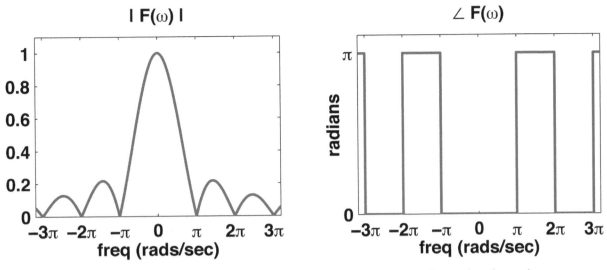

Figure 3-7: Fourier Transform of square pulse x(t) using a = 1. Left: Magnitude plot; Right: Phase plot.

The resulting plot is shown in Figure 3-7. Note two things. First, look at the zero crossings of the magnitude function. These are at integer multiples of $\omega = \pi$ rads/sec. This is expected since the Fourier Transform is $F(j\omega) = \sin(\omega)/\omega$. The zero crossings will be at the frequencies that make $\sin(\omega)$ equal to zero, namely integer multiples of π, or $\omega = n\pi$ where n is all non-zero integers.

The second thing worth noticing is the phase plot. Note that the phase is always either zero or π radians. This is as expected, because our Fourier Transform is purely real. Numbers that are purely real can have only one of two phases: zero (if the number is positive) and π (if the number is negative). For example, consider the number −0.2. Expressed in polar form, this number can be written as $0.2e^{j\pi}$. Therefore its magnitude is 0.2 and its phase is π radians. Purely real numbers must always have phase angles of either zero or π radians.

Let's try some variations on our original signal x(t) and see if we can determine the Fourier Transforms without actually doing any calculus. Consider the two signals shown below in Figure 3-8.

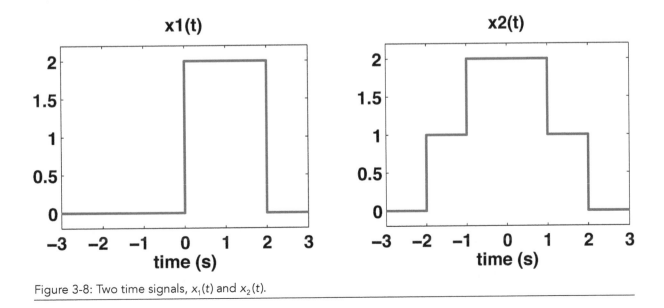

Figure 3-8: Two time signals, $x_1(t)$ and $x_2(t)$.

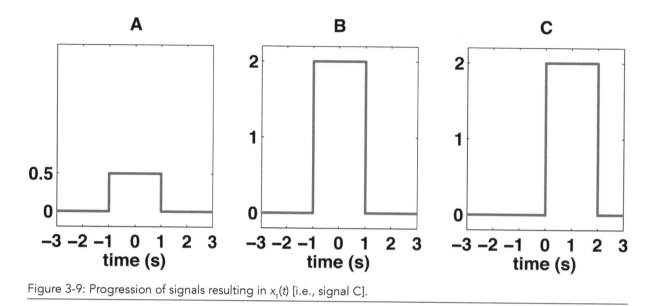

Figure 3-9: Progression of signals resulting in $x_1(t)$ [i.e., signal C].

The first signal is a square pulse that has been shifted and scaled. This concept is explained by the progression of signals shown in Figure 3-9.

In Figure 3-9, the first signal, A, is a square pulse with $a = 1$ and an area of 1. We can therefore find the Fourier Transform of signal A by applying Equation [3-4] with $a = 1$: $F(j\omega) = \sin(\omega)/\omega$. Next we consider signal B. The only difference between signals A and B is that B is four times the height of A. Therefore, we can find the Fourier Transform of B by multiplying the Fourier Transform of A by 4: $F(j\omega) = 4\sin(\omega)/\omega$. Finally we compare signals B and C. The only difference is that C has been shifted to the right by one second relative to B. We learned in Section 2.4.4 that applying a time shift to the right of τ seconds is equivalent to multiplying the Fourier Transform by $e^{-j\omega\tau}$. Therefore, the Fourier Transform of signal $x_1(t)$ [i.e., signal C] is $F(j\omega) = 4\sin(\omega)e^{-j\omega}/\omega$. Of course if you were so inclined, you could arrive at the very same answer by doing the Fourier integral from scratch.

Next, let's find the Fourier Transform of signal $x_2(t)$ from Figure 3-8. There are a couple of ways we could solve this, but the easiest is to notice that $x_2(t)$ can be expressed as a sum of two square pulses, as shown in Figure 3-10.

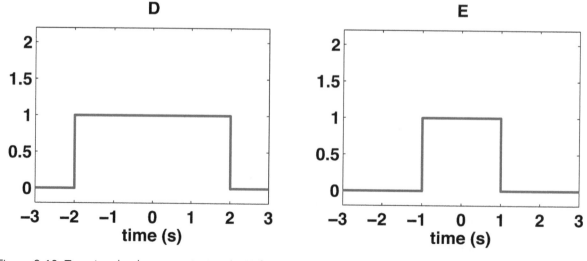

Figure 3-10: Two signals whose sum is signal $x_2(t)$ from Figure 3-8.

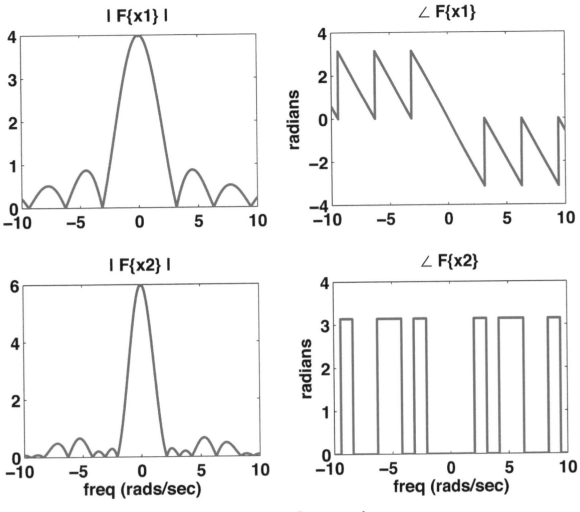

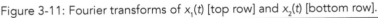

Figure 3-11: Fourier transforms of $x_1(t)$ [top row] and $x_2(t)$ [bottom row].

These are two square pulses; the first has $a = 2$ and a multiplier of 4, and the second has $a = 1$ and a multiplier of 2. The Fourier Transform of $x_2(t)$ is therefore $F(j\omega) = 4\sin(2\omega)/2\omega + 2\sin(\omega)/\omega$. Not bad! No calculus and we easily arrive at the right answer. Again, you can get the exact same answer by applying the Fourier Transform integral directly to the signal $x_2(t)$.

The Fourier Transforms of $x_1(t)$ and $x_2(t)$ are shown in Figure 3-11. By comparing the magnitude plots, we see that signal $x_2(t)$ as its energy more concentrated at low frequencies as compared to $x_1(t)$. This makes intuitive sense. Signal $x_1(t)$ has tall sharp edges (located at $t = 0$ and $t = 2$ seconds). We've learned that sharp edges are high-frequency signals. By comparison, $x_2(t)$ has stair-step edges, which allow the signal to increase more gradually. This gradual increase requires lower-frequency components to accomplish. Therefore we expect that the magnitude response of $x_2(t)$ shows energy clustered at lower frequencies, where as $x_1(t)$ has more of its energy at higher frequencies.

Section 3.3 Impulses and Constants

In the previous section, we learned how to take the Fourier Transform of a square pulse. Now we'll focus on two special cases, the constant signal and the impulse signal.

3.3.1 Square Pulse

Previously, we learned that the transform of a square pulse $x(t)$ that spans $-a \leq t < a$ with a height of $1/2a$ (i.e., area is one) is $F(j\omega) = \sin(a\omega)/a\omega$. Let's examine what happens to the Fourier Transform as we vary the value of a, and why this makes sense intuitively. Figure 3-12 shows two square pulses, $x_1(t)$ and $x_2(t)$, and their corresponding magnitude Fourier Transform plots (we omit the phase plots for simplicity; we can make our point without them).

Note that $x_1(t)$ has $a = 1$ while $x_2(t)$ has $a = 5$. Now consider the corresponding Fourier magnitude plots. For $x_1(t)$, the zero crossings are at integer multiples of π, whereas for $x_2(t)$, the zeros are at integer multiples of $\pi/5$. This is an interesting phenomenon. It appears that as our square pulse gets stretched out in time (i.e., a gets larger), the frequency representation gets squashed down toward the origin (i.e., zero crossings get closer to each other and closer to the origin). Let's think why this makes sense intuitively. As the square pulse gets stretched out, we get more of the flat part (i.e., in the range $-a \leq t < a$), which is the *low-frequency* part of the signal. Recall that signals that don't change very quickly (like our flat bit) are low frequency whereas signals that do change quickly (like the edges of the square pulse) are high frequency. So if we get more

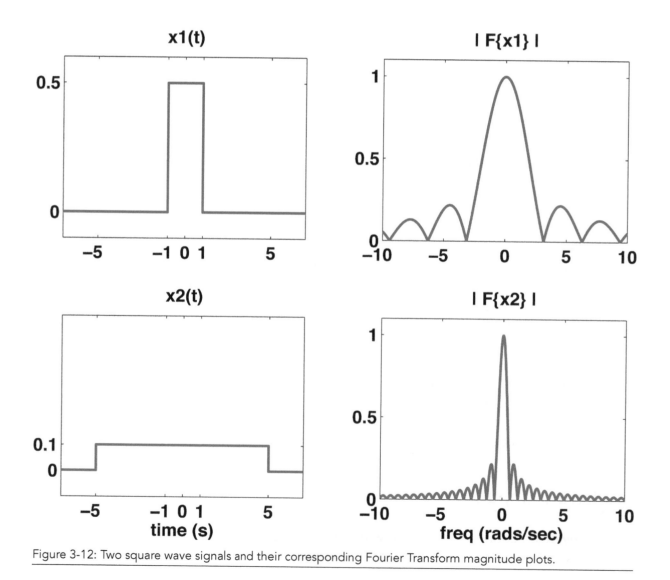

Figure 3-12: Two square wave signals and their corresponding Fourier Transform magnitude plots.

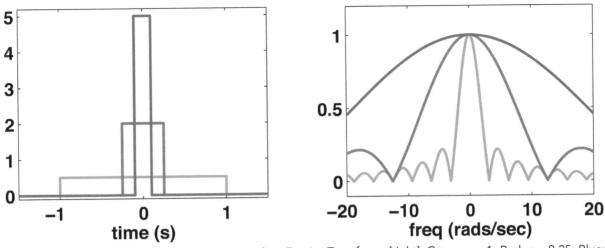

Figure 3-13: Square pulses [left] and corresponding Fourier Transforms [right]. Green: a = 1; Red: a = 0.25; Blue: a = 0.1.

low-frequency signal as we stretch out our square pulse, it stands to reason that the Fourier Transform should reflect this fact. Indeed, as we can see in Figure 3-12, the signal's energy becomes more heavily concentrated at low frequencies when the square pulse is stretched out. In general, the point we are trying to make here is pretty important and will arise over and over in signal processing theory: the broader a signal becomes in the time domain, the narrower it will become in the frequency domain, and vice versa.

3.3.2 Impulse

Now let's look at our first special case of the square pulse. Suppose we started to make our square pulse narrower and narrower. In other words, suppose we start decreasing a. What happens to the Fourier Transform? Figure 3-13 gives us an idea.

From Figure 3-13 we see that, as our pulse gets narrower in time, it gets broader in frequency. In fact, we can imagine what might happen in the limit as $a \to 0$. In that case, the square pulse would become an *impulse*, denoted by $\delta(t)$. An impulse is something that has zero width and infinite height but finite area. In our case, the area of the impulse would be one. In the frequency domain, as $a \to 0$, the Fourier Transform would become so wide that it would just become the constant $F(j\omega) = 1$. This is a very important and common Fourier Transform. It tells us that an impulse consists of equal amounts of all frequencies, even out to infinity. In other words, an impulse is the ultimate "broadband" signal.

3.3.3 Constant

In the previous section, we saw what happened to the square pulse as $a \to 0$. Now, instead of making a smaller, we'll make it bigger and see what happens to the Fourier Transform. One small change we'll need to make for this example is that instead of our square pulse having a constant area of one, we will make it so that it has a constant *height* of one. It is not hard to show that when $x(t) = 1$ in the range of $-a \le t < a$, then the Fourier Transform is given by $F(j\omega) = \sin(a\omega)/\omega$. Figure 3-14 shows the signal $x(t)$ and its Transform for several values of a.

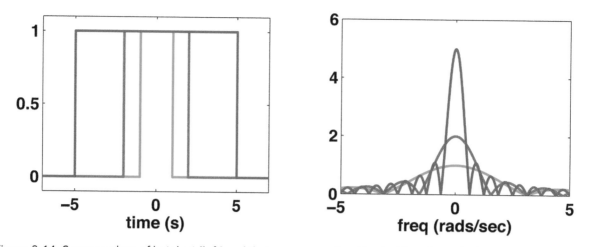

Figure 3-14: Square pulses of height 1 [left] and their corresponding Fourier Transforms [right]. Green: a = 1; Red: a = 2; Blue: a = 5.

We note that as a gets bigger (i.e., the square gets wider), then the Fourier Transform gets narrower and taller. In the limit as $a \to \infty$, we can infer that the signal $x(t)$ will just become a constant in time and the Fourier Transform will be come an impulse at $\omega = 0$ rads/sec in the frequency domain: $F(j\omega) = \delta(j\omega)$. This is interesting for two reasons. First, it tells us that a constant signal in time is composed of a single frequency: $\omega = 0$. There is no other energy at any other frequency whatsoever. This should make some amount of intuitive sense, since we've already discussed that a constant function is indistinguishable from a cosine with frequency $\omega = 0$ rads/sec. The second reason this is interesting is that this is essentially the converse of the conclusion we discovered in the previous section. Whereas the Fourier Transform of an impulse in time is a constant in frequency, it is also true that the Fourier Transform of a constant in time is an impulse in frequency! This converse relationship between signals in the time and frequency domain is a common theme that we will see again in the future.

For completeness, we show in Figure 3-15 and Figure 3-16, the Fourier Transforms of an impulse function and of a constant function, respectively.

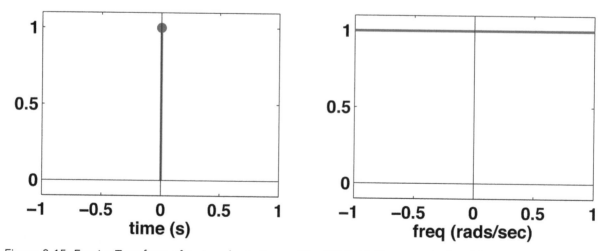

Figure 3-15: Fourier Transform of an impulse in time: $x(t) = \delta(t)$ [left]. The magnitude of the Fourier Transform is simply the constant 1 [right]. The phase plot is omitted because it is zero. This plot demonstrates that an impulse contains energy at all frequencies equally, and is therefore the ultimate broadband signal.

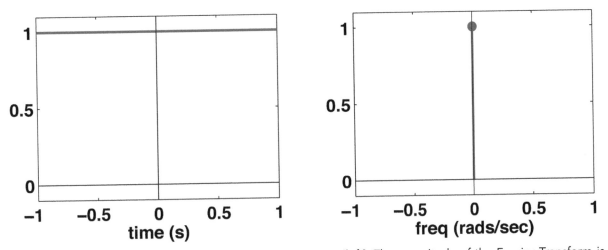

Figure 3-16: Fourier Transform of a constant in time: $x(t) = 1$ [left]. The magnitude of the Fourier Transform is simply the impulse $F(j\omega) = \delta(j\omega)$ [right]. The phase plot is omitted because it is zero. This plot demonstrates that a constant signal contains energy only at frequency $\omega = 0$. Constant functions are therefore the ultimate narrow-band signal.

Section 3.4 Plotting Magnitude and Phase

In this section, we will practice some of the skills necessary for plotting the magnitude and phase of the Fourier Transforms of square pulses. We will create the plots "by hand" and then again using Matlab. A good deal of this was covered in Section 3.2 but here we delve deeper. We start by considering the plot shown in Figure 3-17. We learned before that the Fourier Transform of this function is

$$F(j\omega) = \frac{\sin(\omega)}{\omega} \qquad [3\text{-}7]$$

Figure 3-17: Square pulse $x(t)$.

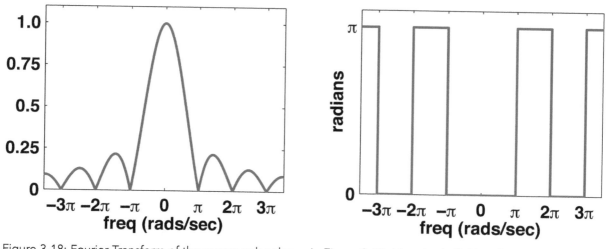

Figure 3-18: Fourier Transform of the square pulse shown in Figure 3-17. Magnitude [left] and Phase [right].

The magnitude and phase plots for $F(j\omega)$ are shown in Figure 3-18. Note that, as expected, the zero-crossings are at integer multiples of π, and that the phase angles are all either zero (indicating that $F(j\omega)$ was a positive real number) or π (indicating that $F(j\omega)$ was a negative real number).

Now let's try finding the Fourier Transform of a time-shifted version of $x(t)$, as shown in Figure 3-19. We know that the effect of shifting a signal to the right by τ seconds is to multiply the Fourier Transform by $e^{-j\omega\tau}$. In our case, there has been a right shift of $\tau = 1$s and therefore

$$F(j\omega) = \frac{\sin(\omega)}{\omega}e^{-j\omega} \qquad [3\text{-}8]$$

We now wish to plot the magnitude and phase of Equation [3-8]. In comparing Equations [3-7] and [3-8], we see that the only difference is that we've multiplied by $e^{-j\omega}$. When we multiply two complex functions, as we have here, their magnitudes become multiplied and their phases become summed. In other words, the magnitude of Equation [3-8] will be the same as that of Equation [3-7] multiplied by the magnitude of $e^{-j\omega}$, which is just one. Therefore, the magnitude of Equation [3-8] will be the same as that of Equation [3-7]. This should make a good deal of intuitive sense—as we've argued before, shifting a signal in time doesn't actually affect its

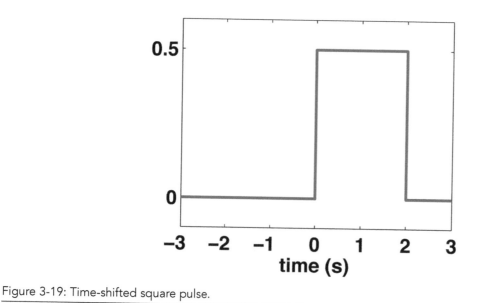

Figure 3-19: Time-shifted square pulse.

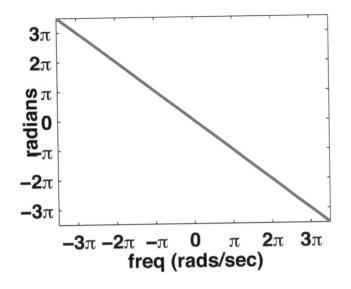

Figure 3-20: Phase plot of $e^{-j\omega}$.

frequency content. It still contains the same amounts of the same frequencies. They will just happen to be located differently. This means that we should expect the magnitude to stay the same, but not the phase.

The phase plot is a bit more complicated. The phase plot for Equation [3-8] should be equal to the sum of the phase from Equation [3-7] (see Figure 3-18) and the phase of $e^{-j\omega}$. The phase of $e^{-j\omega}$ is simply $-\omega$, and its plot is shown in Figure 3-20.

Summing together the phase plot of Figure 3-18 with Figure 3-20, we arrive at the phase plot of Equation [3-8].

We are pleased to see that the phase plot is an odd function, which will always be the case as long as $x(t)$ is purely real (which will always be the case this semester).

Finally, let's try using Matlab to plot our function. The code for creating both magnitude and phase plots is as follows.

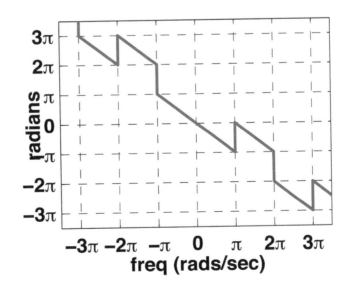

Figure 3-21: Phase plot of Equation [3-8].

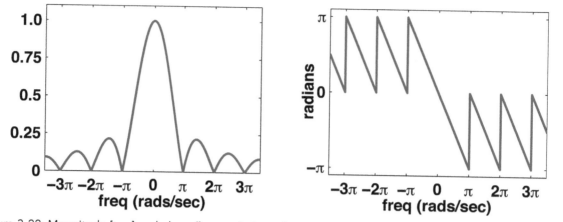

Figure 3-22: Magnitude [top] and phase [bottom] plots of Equation [3-8].

```
w = linspace(-3.5*pi,3.5*pi,1e5);
F = sinc(w/pi).*exp(-j*w);

subplot(1,2,1);
plot(w,abs(F));
xlabel('freq (rads/sec)');

subplot(1,2,2);
plot(w,angle(F));
xlabel('freq (rads/sec)');
```

The resulting plots are shown in Figure 3-22.

Happily, the magnitude plot is precisely the same as in Figure 3-18, which is as expected. The phase plot, however, is slightly different than our "hand calculation" from Figure 3-21. Fortunately, there is a simple explanation. If we look at Figure 3-21, we see that the predicted phase at $\omega = 2\pi$ rads/sec is -2π rads. Recall, however, that for complex numbers such as $Me^{j\phi}$, the phase φ is periodic. In other words, the phase angle of 0 rads is the same as 2π rads and 4π rads. Keeping this in mind, we see that the phase angle of -2π rads predicted from Figure 3-21 is equivalent to 0 rads, which is precisely what is seen in Figure 3-22. In fact, the phase plots of Figure 3-21 and Figure 3-22 are exactly the same once we take into account the "wrap-around" property of the phase angle of complex numbers.

Section 3.5 Properties of the Fourier Transform

The Fourier Transform has many convenient properties that we can use to help us manipulate and understand signals in the frequency domain. Many of the properties derived in Section 2.4 for the Fourier Series also apply for the Fourier Transform. In this section we'll focus on properties that work a little differently in the Fourier Transform.

3.5.1 Time-Scaling Property

In Section 2.4.6, we learned that compressing a signal in time did not affect the Fourier Series coefficients; the only thing that changed was the values of the harmonic frequencies. The time-scaling property works a little differently for aperiodic signals. Starting with the assumption that signal $x(t)$ has Fourier Transform $X(\omega) = \int x(t)e^{-j\omega t}dt$, let's see if we can derive the Fourier Transform of $y(t) = x(at)$, which is a time-scaled version of $x(t)$. Remember, time scaling just means that we are stretching or compressing the signal along the x-axis but otherwise not changing its shape in any way.

$$Y(j\omega) = \int y(t)e^{-j\omega t}dt$$

$$= \int x(at)e^{-j\omega t}dt$$

Let $u = at$ and $du = a \cdot dt$

$$= \int x(u)e^{-\frac{j\omega u}{a}}\frac{du}{a}$$

$$= \frac{1}{a}\int x(u)e^{-j\frac{\omega}{a}u}du$$

$$= \frac{1}{a}X\left(\frac{\omega}{a}\right) \qquad\qquad [3\text{-}9]$$

This is a very interesting result. It tells us that if we compress the signal in time by a factor of a, then the only effects on the Fourier Transform are that we stretch it out by that same factor of a and that we divide its magnitude down by a as well. As always, this makes intuitive sense. If you compress a signal in time, then you are reducing its low-frequency components and increasing its high-frequency components. This in turn means that you would expect the Fourier Transform to stay mostly the same except you will "stretch" it out such that there is more energy at high frequencies. This is exactly the trend shown in Figure 3-14.

Let's practice applying Equation [3-9]. We start with a square pulse with $a = 1$ and a height of 2 (i.e., the green trace from Figure 3-14). By inspection, we know the Fourier Transform must be $X_{green}(j\omega) = 2\sin(\omega)/\omega$. Now we stretch the signal in time by a factor of five to get the blue trace from Figure 3-14. Mathematically, we are saying that $x_{blue}(t) = x_{green}(t/5)$. Equation [3-9] tells us that the Fourier Transform must be $X_{blue}(j\omega) = 5X_{green}(5j\omega) = 10\sin(5\omega)/(5\omega)$. Happily, this is exactly the answer that we would have gotten by inspection.

3.5.2 Derivatives and Integrals

In this section, we'll learn what happens to the Fourier Transform of a signal when we take its integral or derivative. We'll also learn how to exploit this property to quickly calculate some common Fourier Transforms without using any calculus. The real benefit of this technique won't become truly apparent, however, until a little later, once we've introduced the concept of systems.

You might recall that we built the Fourier Transform around the idea that any time-based signal can be rebuilt as a sum of cosines, provided we pick the right amplitude, phase, and frequencies of cosines. Recall also that each of those cosines can be expressed as

$$x(t) = Ae^{j\omega t} + A^* e^{-j\omega t} \qquad [3\text{-}10]$$

where the amplitude and phase of the cosine are $2|A|$ and $\angle A$, respectively.

Now suppose we take the derivative of $x(t)$. Using the chain rule, we find

$$x'(t) = (j\omega)Ae^{j\omega t} + (-j\omega)A^* e^{-j\omega t} \qquad [3\text{-}11]$$

This is quite interesting. This equation tells us that if we take the derivative of $x(t)$, the Fourier Transform at frequency ω is equal to $j\omega A$. This means that in general, if we know the FT of $x(t)$, then we can compute the FT of the derivative $x'(t)$ by simply multiplying the FT by $j\omega$. Note that $j\omega$ has a magnitude of ω and a phase of $\pi/2$. This tells us that, from a frequency perspective, taking the derivative means emphasizing the higher frequencies (since the FT of higher frequencies will get magnified by higher values of ω), and conversely de-emphasizing low-frequency signal components. For example, consider the plot shown in Figure 3-23(A).

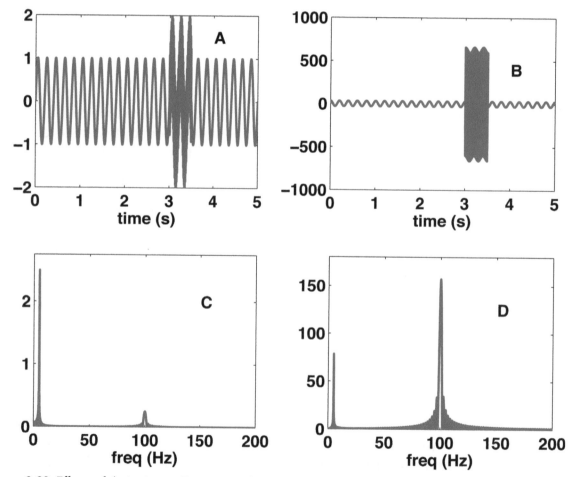

Figure 3-23: Effects of derivative on Fourier Transform. (A) time-signal composed of 5Hz cosine with a short burst of 100Hz cosine. (B) The time derivative of the signal in A. (C) The Fourier Transform of the signal in A, clearly showing peaks at 5 and 100Hz. (D) The Fourier Transform of the signal in B.

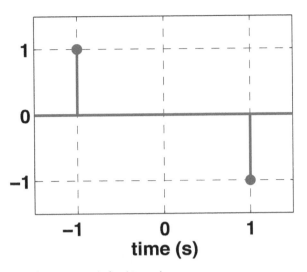

Figure 3-24: Function consisting of two time-shifted impulses.

That signal (A) contains sinusoids at both 5 Hz and 100 Hz; both these peaks can clearly be seen in the Fourier Transform of A, seen in Figure 3-23(C). In Figure 3-23(B), we see the derivative of signal A. The derivative clearly shows that the high-frequency burst has been emphasized in terms of magnitude; Figure 3-23(D) illustrates the same phenomenon in the frequency domain. Our observation is therefore consistent with the theory, as predicted in Equation [3-11]: *derivatives* tend to emphasize *higher* frequencies.

By very similar arguments, it can also be shown that integrating a function $x(t)$ means dividing its Fourier Transform by $j\omega$. This means that integrating a function tends to emphasize its lower frequencies and de-emphasize its higher ones.

In summary, suppose we know that the Fourier Transform of $x(t)$ is $F(j\omega)$. We can then say that

$$x'(t) \Leftrightarrow (j\omega)F(j\omega)$$

$$\int x(t)dt \Leftrightarrow F(j\omega)/(j\omega) \tag{3-12}$$

This turns out to be a really useful tool. Recall from Section 3.3.2 that the Fourier Transform of an impulse $x(t) = \delta(t)$ is simply $F(j\omega) = 1$. We can use this property, in conjunction with the time-shift property, to quickly write down the Fourier Transform of the function shown in Figure 3-24. By inspection, we can say that the Fourier Transform is given by $F(j\omega) = 1e^{j\omega} - 1e^{-j\omega}$. Recall that shifting something (in this case, an impulse) to the left by τ units means multiplying the Fourier Transform by $e^{j\omega\tau}$; shifting to the right by τ units means multiplying by $e^{-j\omega\tau}$.

Next, try to imagine what the integral of the signal in Figure 3-24 might look like. Hopefully, it should be fairly apparent that the integral of Figure 3-24 will be the signal shown in Figure 3-25. Finally we can put all the pieces together: if we know that the Fourier Transform of Figure 3-24 is $F(j\omega) = e^{j\omega} - e^{-j\omega}$, and we know that Figure 3-25 is the integral of Figure 3-24, then we can simply say that the Fourier Transform of Figure 3-25 is $F(j\omega) = (e^{j\omega} - e^{-j\omega})/j\omega = 2\sin(\omega)/\omega$. Is this the right answer? We can solve for the Fourier Transform of Figure 3-25 in at least two other ways! One is to simply apply the Fourier Transform integral as shown here.

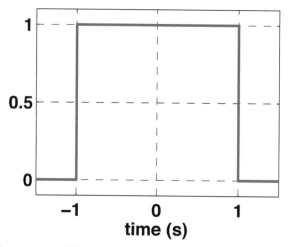

Figure 3-25: This function is the integral of the function shown in Figure 3-24.

$$F(j\omega) = \int_{-1}^{1} 1e^{-j\omega t} \, dt$$

$$= \frac{1}{-j\omega} \left[e^{-j\omega t} \right]_{-1}^{1}$$

$$= \frac{1}{-j\omega} \left[e^{-j\omega} - e^{j\omega} \right]$$

$$= \frac{2\sin(\omega)}{\omega} \qquad\qquad [3\text{-}13]$$

The other method is to just solve by inspection along the lines of the examples in Section 3.2. All three methods yield the same answer.

Next, we'll do two examples to see how the derivative and integral properties can quickly get us a Fourier Transform without too much trouble.

Example 3-1

Consider the signal $x(t)$ shown in Figure 3-26. Our goal is to solve for its Fourier Transform with the least amount of work possible.

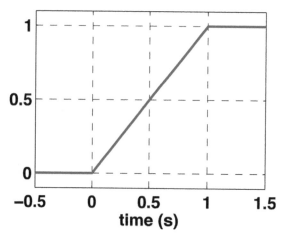

Figure 3-26: Time signal, $x(t)$.

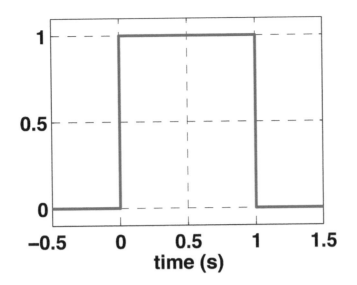

Figure 3-27: $x'(t)$, the derivative of $x(t)$.

This function is defined as

$$x(t) = \begin{cases} 0 & t < 0 \\ t & 0 \le t < 1 \\ 1 & t > 1 \end{cases} \qquad [3\text{-}14]$$

The "hard" way to solve this Fourier Transform will be to apply the actual integral: we'll do that later. For now, let's start by taking the derivative of $x(t)$, as shown in Figure 3-27.

We learned in Section 3.2 how to compute the Fourier Transform of a square pulse by inspection. In this case, the FT of $x'(t)$ is

$$F_{deriv}(j\omega) = \frac{\sin(\omega/2)}{\omega/2} e^{-j\omega/2} \qquad [3\text{-}15]$$

Finally, to arrive at the FT of $x(t)$, we simply "integrate" the FT of $x'(t)$, which means dividing $F_{deriv}(j\omega)$ by $j\omega$.

$$F(j\omega) = \frac{\sin(\omega/2)}{(\omega/2)j\omega} e^{-\frac{j\omega}{2}}$$

$$= \frac{2\sin(\omega/2)}{j\omega^2} e^{-\frac{j\omega}{2}} \qquad [3\text{-}16]$$

That was relatively painless. For comparison, let's try arriving at the same answer using the more conventional integral approach.

$$F(j\omega) = \int_0^1 t\, e^{-j\omega t} dt + \int_1^\infty 1\, e^{-j\omega t} dt$$

$$= \left[\frac{e^{-j\omega t}}{\omega^2}(j\omega t + 1) \right]_0^1 + \frac{1}{-j\omega}\left[e^{-j\omega t} \right]_1^\infty$$

$$= \frac{1}{\omega^2}\left[e^{-j\omega}(j\omega+1)-1\right]-\frac{1}{j\omega}\left(0-e^{-j\omega}\right)$$

$$= \frac{1}{j\omega^2}\left(-\omega e^{-j\omega}+je^{-j\omega}-j+\omega e^{-j\omega}\right)$$

$$= \frac{1}{j\omega^2}\left[je^{-\frac{j\omega}{2}}\left(e^{-\frac{j\omega}{2}}-e^{\frac{j\omega}{2}}\right)\right]$$

$$= \frac{2\sin(\omega/2)}{j\omega^2}e^{-\frac{j\omega}{2}} \qquad\qquad\qquad [3\text{-}17]$$

Phew! That was a lot of pointless calculus, especially considering the much easier approach set forth in Equation [3-16]. For "fun," it might be interesting to compare the Fourier Transforms of $x(t)$ and $x'(t)$, to see whether or not our intuition about derivatives and Fourier Transforms holds in this case. The plots are shown in Figure 3-28.

Figure 3-28 shows a couple of interesting things. First off, our intuition from Section 3.5.2 was that taking the derivative means emphasizing (or adding more energy to) the higher frequencies. This is exactly what we observe in this case. Comparing the FT of $x(t)$ [left] versus $x'(t)$ [right], we see that the derivative signal (at right) has more energy at higher frequencies than the original signal (at left) does.

The other interesting thing to observe is the value of both Fourier Transforms at $\omega = 0$ rads/sec. You should remember that the value of the Fourier Transform at $\omega = 0$ always equals the DC offset of the signal. Or equivalently, it tells us the value of the integral of the signal over all time.

$$F(j\omega = 0) = \int x(t)e^{-j0t}dt$$

$$= \int x(t)dt \qquad\qquad\qquad [3\text{-}18]$$

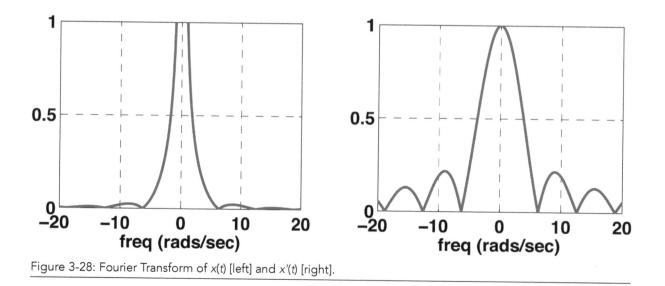

Figure 3-28: Fourier Transform of $x(t)$ [left] and $x'(t)$ [right].

For the original signal $x(t)$, we see that its integral over all time is infinity. Therefore we are not surprised to see in Figure 3-28 [left] that the magnitude of the Fourier Transform at $\omega = 0$ is also infinity. Likewise, we can see that the integral of $x'(t)$ [see Figure 3-27] is just 1. Therefore, it makes sense that the magnitude of the Fourier Transform at $\omega = 0$ [Figure 3-28, right] also equals 1.

Example 3-2

The second example problem will be left as an exercise to the reader (with a nice hint, of course). Solve for the Fourier Transform of the signal $x(t)$ shown in Figure 3-29.

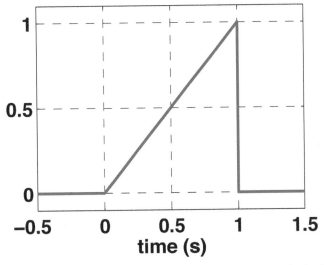

Figure 3-29: Signal $x(t)$. Can you take its Fourier Transform without doing any calculus?

The hint is that the signal can be represented as the sum of two other signals (see Figure 3-30). The Fourier Transform for both of these other two signals can be found easily enough using the derivative/integral properties of the Fourier Transform.

Just so you can check your work, the final answer should be

$$F(j\omega) = \frac{j\omega e^{-j\omega} + e^{-j\omega} - 1}{\omega^2}$$ [3-19]

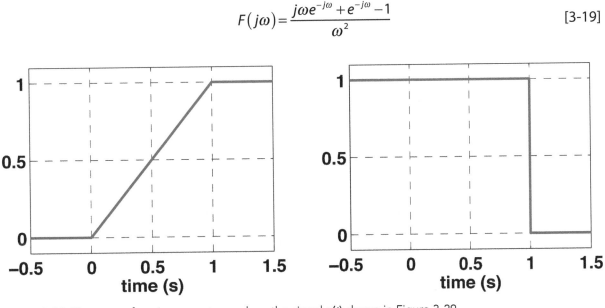

Figure 3-30: These two functions sum to produce the signal $x(t)$ shown in Figure 3-29.

Section 3.6 Fast Fourier Transform

So far we've learned how to apply the Fourier Transform integral to find the frequency content of continuous-time signals that can be expressed as functions of time. However, in real life, many interesting signals cannot be cleanly expressed as a function. For example, a recording of someone's voice can't be expressed as a function, but we might still be very interested in understanding its frequency content.

Fortunately, there are methods available for numerically computing the Fourier Transform of *any* arbitrary digital signal. The best-known such method is the Fast Fourier Transform, or FFT for short. The FFT is a function that can compute the frequency content of any signal $x(t)$. For example, consider the signal $x(t) = cos(2\pi t)$, as shown in Figure 3-31. We know that this is a "pure-tone" signal composed of energy at exactly one frequency, 1 Hz. Let's see if we can use Matlab's built-in FFT command to validate that this is the case. First, create the signal $x(t)$. This requires first defining its time vector. Note that for this example, we're going to need to precisely define the sampling rate fs in order to make a meaningful frequency plot.

```
fs = 1000;
dt = 1/fs;
t = 0:dt:10;
x = cos(2*pi*t);
```

Next we take the FFT of $x(t)$.

```
X = fft(x);
```

The vector X will have the same number of elements as the signal vector x. However X will be a vector of complex numbers, since the Fourier Transform is a complex function.

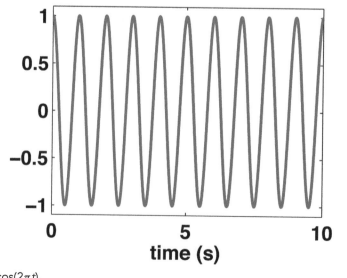

Figure 3-31: Signal $x(t) = cos(2\pi t)$.

Before we plot the FFT, let's do two more manipulations. The first simply reshuffles the order of the values in the vector X so that the frequency 0Hz appears right in the middle.

```
X = fftshift(X);
```

The second manipulation is to scale the values of the FFT by multiplying by dt. The reasons for this are a bit arcane and beyond the scope of this book, but unless we do this step, the values of the FFT won't work out to the values we'd expect from the hand integrals.

```
X = X*dt;
```

Finally we need to create a frequency vector to plot against X. This is fairly straightforward.

```
f = linspace(-fs/2,fs/2,length(X));
```

In this case we have chosen to create a frequency vector in Hertz but we could just as easily create a frequency vector in rads/sec by multiplying f by 2π.

```
w = 2*pi*f;
```

Finally, let's put all the pieces together and plot the magnitude of the FFT (remember that FFT returns complex values and therefore we have to plot either the magnitude or the phase, or both). The resulting plot is shown in Figure 3-32.

```
X = fftshift(fft(x))*dt;
f = linspace( -fs/2 , fs/2 , length(X) );
plot(f,abs(X));
```

Figure 3-32 [left] clearly shows the expected peaks at ± 1Hz. Also note that the magnitude plot is even, meaning that it is a mirror image across the y-axis. We have examined this property

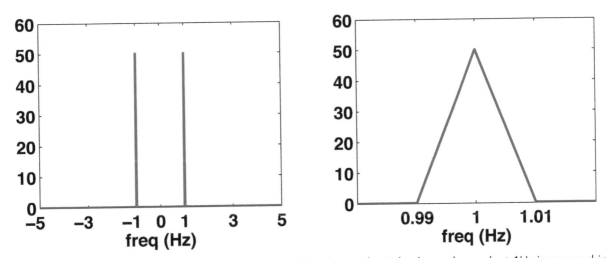

Figure 3-32: FFT for the signal shown in Figure 3-31. The plot on the right shows the peak at 1Hz in zoomed-in detail.

of the Fourier Transform earlier and therefore we shouldn't be surprised to see it again here. However, if we know that the FFT will *always* have this even property, there is no real reason to plot the negative frequencies, since we implicitly know what they will be if we have properly plotted the positive frequencies. For this reason, from here on, the FFT will be plotted only over positive values of frequency.

The right-hand plot in Figure 3-32 shows zoomed-in detail of the 1Hz peak. Let's analyze it to see if it makes sense. We know that since $\cos(2\pi t) = \frac{1}{2}e^{j2\pi t} + \frac{1}{2}e^{-j2\pi t}$, we expect the Fourier Transform to be a pair of impulses located at $\pm 2\pi$ rads/sec $= \pm 1$ Hz, and the area of each of those impulses is 1/2. The plot in Figure 3-32 shows a triangle with area $0.5 \times 50 \times 0.02 = 0.5$. Therefore, the FFT has correctly predicted impulses at ± 1 Hz with area 0.5. We can now use the FFT function with confidence to analyze any arbitrary signal.

3.6.1 Clipped Cosine Signal

The FFT function appears to be quite useful. Let's apply it to a couple more cases to see what we can learn. Consider the signal shown in Figure 3-33.

This signal is the same as $x(t)$ from Figure 3-31 except that it has been *clipped*. This means that we have sliced off the tops and bottoms of our sine wave, in this case at ± 0.5. This is an example of what might happen to an audio signal in a poorly designed amplifier if the user attempts to amplify the signal above what the amplifier can handle. The natural question to ask is whether or not clipping the signal changes the frequency content of the signal. We apply our FFT operator and plot the result, shown in Figure 3-34.

Here we see that the clipped signal has different frequency elements than the original pure-tone cosine. If we compare Figure 3-32 and Figure 3-34, we see that the clipped signal has *harmonics* at integer multiples of 1Hz. Figure 3-34 clearly shows harmonics at $f = 3, 5, 7,$ and 9 Hz. In a way, this isn't surprising since the clipped sine wave looks somewhat like a square wave signal, which we learned long ago has energy at all the odd harmonics.

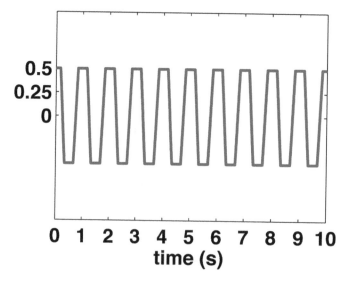

Figure 3-33: Clipped cosine wave.

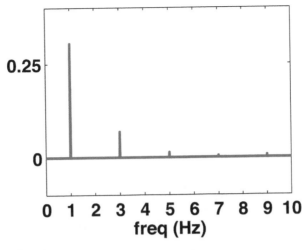

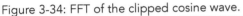

Figure 3-34: FFT of the clipped cosine wave.

The point of this example is two-fold. First, the FFT function provided a quick and painless way to examine the frequency content of the clipped cosine signal. Solving for the Fourier Transform by hand would have been a complete headache. The second point is that the simple act of clipping the signal added energy at a bunch of frequencies that weren't present in the original signal. If we were listening to music and this happened, our clipped music signal would sound different (and probably much worse) since it would contain a number of frequencies that the musician never intended for you to hear. This is an example of how frequency analysis can be very valuable to an engineer.

3.6.2 ECG Signal

Our final example is ECG data from a human heart (Figure 3-35). We see from the figure that the heart beats approximately 1.6 times every second (five beats in three seconds). We also see that each heartbeat is marked by a conspicuous upstroke (the ventricle beat) as well

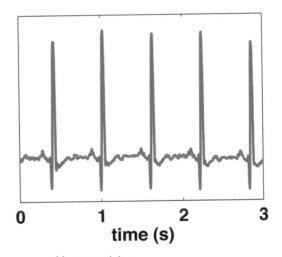

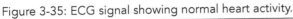

Figure 3-35: ECG signal showing normal heart activity.

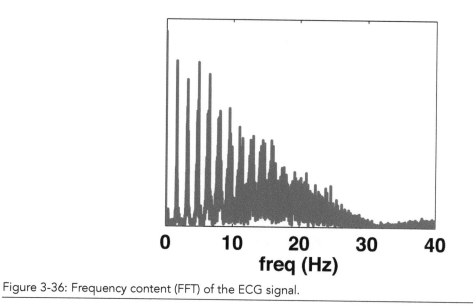

Figure 3-36: Frequency content (FFT) of the ECG signal.

Table 3-1: Summary of common Fourier transform pairs.

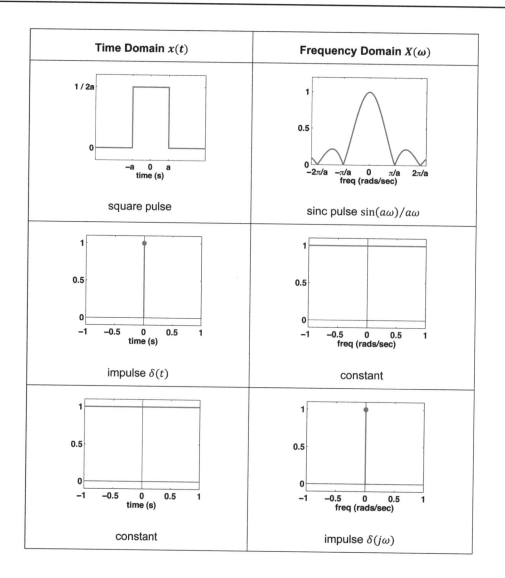

as a smaller upstroke immediately prior (the atrial beat) and a little dimple following (the repolarization). As signal engineers, however, we also want to know how the signal looks as a function of frequency. We apply the FFT and arrive at Figure 3-36.

The first interesting thing we notice is that the signal is composed of a series of spikes. These look an awful lot like harmonics, and they are. A closer look shows that the first harmonic occurs at f = 1.6 Hz. This is number *exactly* matches the 1.6 beats per second we calculated by eyeballing Figure 3-35. This is no coincidence! The harmonics are at integer multiples of 1.6 Hz and collectively they give the ECG signal its particular shape.

The other thing worth noticing from Figure 3-36 is that there is hardly any energy in the signal above 30 Hz. This knowledge will become especially valuable down the line when we start discussing sampling theory.

Section 3.7 Summary

This chapter has introduced the concept of the Fourier Transform, which converts a signal from a time representation to its frequency representation. It tells us how much energy a signal contains and various frequencies. This chapter has also derived the Fourier Transform of some common signals, presented an intuitive view of the Transform, and extended the list of important Transform properties.

The Fourier Transform of some signal x(t) is given by

$$F(j\omega) = \int_{-\infty}^{\infty} x(t)e^{-j\omega t}\, dt \qquad\qquad [3\text{-}20]$$

Some common Fourier Transform signal pairs are shown in Table 3-1. Some useful properties of the Fourier Transform are listed in Table 3-2. We assume that the Fourier Transform of some time signal x(t) is x(ω).

Table 3-2: Common properties of the Fourier transform

Scaling	$cx(t) \Leftrightarrow cX(j\omega)$
Summation	$x_a(t) + x_b(t) \Leftrightarrow X_a(j\omega) + X_b(j\omega)$
Time Shift	$x(t-\tau) \Leftrightarrow e^{-j\omega\tau} X(j\omega)$
Even Functions	$X(j\omega)$ is purely real
Odd Functions	$X(j\omega)$ is purely imaginary
Time Scaling	$x(at) \Leftrightarrow \dfrac{1}{a} X\left(\dfrac{j\omega}{a}\right)$
Derivative	$x'(t) \Leftrightarrow (j\omega) X(j\omega)$
Integral	$\int x(t) \Leftrightarrow \dfrac{1}{j\omega} X(j\omega)$
DC Offset	$\int x(t)\, dt \Leftrightarrow X(j\omega = 0)$

Chapter 4

Systems

So far we've been focused on understanding how to analyze signals in the frequency domain. We've developed the Fourier Transform as a tool for taking a time signal and telling us how to express it as a sum of cosines of different frequencies. We've also learned what high-frequency signals and low-frequency signals look like in various contexts.

In this chapter, we introduce a new concept: the *system*. A system is something that acts upon a signal. For example, a circuit like a voltage divider or an amplifier is a system that acts to modify a voltage signal.

Although there are many ways of interpreting and analyzing systems, we will be focusing on how to analyze them from the perspective of *frequency*. For example, we might like to know whether a given system removes or amplifies certain frequencies from a signal. This analysis fits perfectly with our Fourier analysis: if you know what frequencies are in a given signal, and you know how a system will act upon those various frequencies, you will have a good idea of what the system output will look like.

Section 4.1   First-Order Systems—Differential Equation and Step Response

The easiest way to introduce system frequency analysis is by example. Consider the RC circuit shown in Figure 4-1.

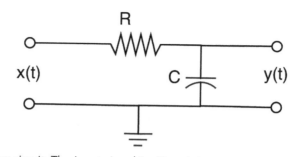

Figure 4-1: Resistor-Capacitor circuit. The input signal is x(t) and the output signal is y(t).

Analyzing this circuit requires deriving and solving its differential equation. We find the differential equation by using Kirchoff's Current Law to equate the current through the resistor to the current through the capacitor. Recall that current through a resistor is equal to the voltage across the resistor divided by the resistance; current through a capacitor is equal to the derivative of the voltage across the capacitor multiplied by the capacitance.

$$I_R = I_C$$

$$\frac{x-y}{R} = C\frac{dy}{dt}$$

$$\frac{dy}{dt} + \frac{y}{RC} = \frac{x}{RC} \qquad \text{[4-1]}$$

Equation [4-1] is a first-order differential equation. If we have an expression for the input signal x(t), we can use Equation [4-1] to determine the output signal y(t), provided we can solve the differential equation. In some cases this is feasible, while in other cases we will have to rely on some Fourier analysis tricks.

In general, all differential equations have two solutions: the homogenous solution and the particular solution. The homogenous solution is the system response in the absence of an input: in other words when x(t) = 0. In a stable system, the homogenous solution will always decay to zero as time goes to infinity. In contrast, particular solution depends on the input signal x(t). We would expect to have a different particular solution for every different input signal. The total solution is just the sum of the homogenous and particular solutions.

Let's start by solving for the homogenous solution of Equation [4-1] by setting x(t) = 0.

$$\frac{dy}{dt} + \frac{y}{RC} = 0$$

$$\frac{dy}{dt} = -\frac{y}{RC}$$

$$\frac{dy}{y} = -\frac{dt}{RC}$$

$$\ln(y) = -\frac{t}{RC} + K$$

$$y_{homogenous} = Ke^{-t/RC} \tag{4-2}$$

We'll solve for the constant K in just a moment after we're done with the particular solution.

To solve for the particular solution, we need to specify an input signal. For now, let's assume that the input signal is a constant, say $x(t) = 1$ for all $t \geq 0$. The easiest way to get the particular solution is to make an educated *guess* and then to verify that this guess works. In our case, it seems reasonable to think that if the input is a constant, then eventually the capacitor will charge up to some voltage and the output will therefore also be a constant, say $y(t) = K_1$. Our job now is to figure out what value K_1 has to be. Assuming $y(t) = K_1$ and $x(t) = 1$, we substitute values into Equation [4-1]. Note that $dy / dt = 0$ because $y(t)$ is a constant.

$$\frac{dy}{dt} + \frac{y}{RC} = \frac{x}{RC}$$

$$0 + \frac{K_1}{RC} = \frac{1}{RC}$$

$$K_1 = 1 \tag{4-3}$$

The particular solution is therefore

$$y_{particular}(t) = 1 \tag{4-4}$$

The complete solution is found by adding Equations [4-2] and [4-4].

$$y(t) = 1 + Ke^{-t/RC} \tag{4-5}$$

Finally, we can solve for the constant K by introducing an initial condition. A perfectly reasonable choice is that $y(t=0) = 0$.

$$y(t=0) = 1 + Ke^{-\frac{0}{RC}}$$

$$0 = 1 + K$$

$$K = -1 \tag{4-6}$$

The complete solution is therefore

$$y(t) = 1 - e^{-t/RC} \tag{4-7}$$

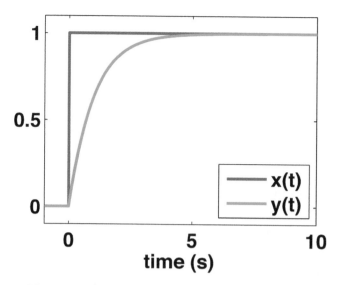

Figure 4-2: Step response to an *RC* circuit. [Blue] Input signal *x(t)*; [Green] Output signal *y(t)*. For simplicity we let *RC* = 1.

Equation [4-7] tells us how the circuit will respond when an input of $x(t) = 1$ is suddenly applied at $t = 0$.

Figure 4-2 shows the input step function $x(t)$ and the corresponding output $y(t)$. We start by analyzing the input signal $x(t)$. We can see that it has a sharp edge, at $t = 0$, and we know from earlier experiences that this sharp edge has to consist of very high frequency signal components. We also see that $x(t)$ is also very flat (i.e., constant) after the step, and that flat (constant) functions consist mostly of very low frequency elements. Therefore, the input signal has both very high and very low frequency elements.

Now consider the output signal $y(t)$. The sharp edge of $x(t)$ has disappeared! We can interpret this to mean that the high frequencies in $x(t)$ that created that sharp edge have been removed by the RC circuit. The output signal $y(t)$ has been stripped of those high-frequency elements. The low-frequency portion of $x(t)$ has been retained, however. We can tell this because eventually $y(t)$ flattens out into a constant (i.e., low frequency) value. To summarize, the RC circuit appears to allow low frequencies to pass through it while rejecting high frequencies. We can therefore describe this circuit as a *low-pass filter*.

The obvious question to ask at this point is, which frequencies are passed and which are rejected? In other words, where do "low" frequencies end and "high" frequencies start? As we will see, the choice of the RC constant is what determines the cutoff between low and high frequencies. For example, can we set RC to reject frequencies above 1 Hz? What if we want to reject only frequencies above 1000 Hz? The relationship between RC and *cutoff frequency* will be explored in this chapter.

Section 4.2 First-Order Systems—Frequency Response

Figure 4-3 shows the RC circuit step response (Equation [4-7]) for three different values of RC: 0.5, 1, and 2. We can see that in all three cases, the RC circuit has behaved as a *low-pass filter*. In all three cases, the high-frequency edge of the step input has been removed and instead, the output signals have a slow curve up to their final value of 1. The low-frequency portion of the input signal (i.e., the flat portion where $x(t) = 1$) is retained in each of the three output signals:

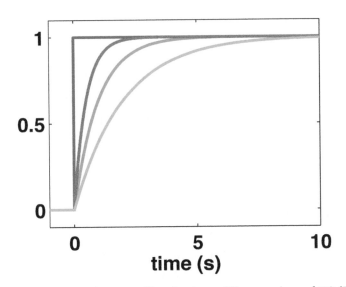

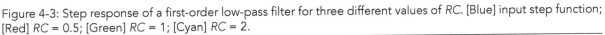

Figure 4-3: Step response of a first-order low-pass filter for three different values of *RC*. [Blue] input step function; [Red] *RC* = 0.5; [Green] *RC* = 1; [Cyan] *RC* = 2.

they all charge up to 1 and then remain there. Therefore in all three cases, low frequencies are retained and high frequencies are rejected.

However, we can also tell that the three filters aren't quite the same. The output of the *RC* = 0.5 filter charges up to its final value of 1 much faster than the other two cases. From a frequency perspective, we can argue that the *RC* = 0.5 filter has allowed more high frequencies to be passed through than the other two filters. Conversely, the *RC* = 2 filter output climbs very very slowly, implying that many more of its high-frequency components have been filtered away. Therefore, the three low-pass filters differ in which low frequencies are retained and which high frequencies are rejected.

Schematically, this concept is captured in Figure 4-4. A low-pass filter can be conceptualized as passing all signals whose frequency is less than some cutoff frequency ω_c, and rejecting all signal frequencies greater than ω_c.

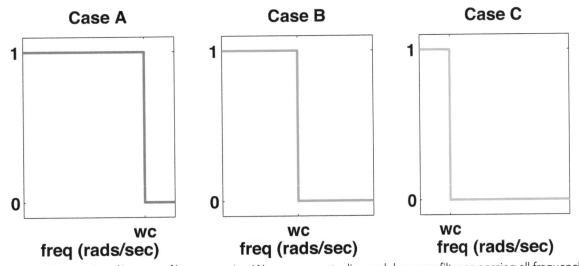

Figure 4-4: Schematics of low-pass filter properties. We can conceptualize each low-pass filter as passing all frequencies less than the cutoff frequency ω_c and rejecting all frequencies greater than ω_c. The only difference between the three low-pass filters is the value of ω_c.

In this schematic, a gain of 1 indicates that all signals are being passed through and a gain of 0 indicates all signals are being rejected. Thinking back to the example shown in Figure 4-3, we can infer that Case A must belong to $RC = 0.5$ since it allows more high-frequency components through. Those high-frequency components contribute to the relatively sharper corner of the step response. Conversely, Case C must correspond to $RC = 2$ since it allows the lowest number of high-frequency elements through. That lack of high frequencies contributes to the very slow corner of the step response. By elimination, Case B must correspond to $RC = 1$.

Therefore we are starting to think about a low-pass filter as being characterized by a cutoff frequency that delineates the boundary between frequencies that are passed and frequencies that are rejected. We have also observed the general relationship that as RC gets larger, the cutoff frequency ω_c gets smaller.

Our eventual goal is to determine the precise relationship between ω_c and RC. We will answer this by solving for the filter output when the input is a cosine: $x(t) = \cos(\omega t)$. If we do our math right, then we should observe a situation where higher-frequency cosines should be suppressed, whereas lower-frequency cosines should be maintained. We seek a precise formula describing this phenomenon.

4.2.1 Cosine Input

Our new challenge is to solve the differential equation in Equation [4-1] when the input signal is $x(t) = \cos(\omega t)$. We will have to employ a few tricks to get the final answer; these will be shown in detail in this section. One thing that will make this process easier is that we can neglect the homogenous solution. Recall that the homogenous solution always goes to zero as time goes to infinity. In our case, we're thinking about the system output in response to a cosine signal of (hypothetically) infinite duration. Therefore the homogenous solution will disappear eventually, meaning we can focus just on the particular solution.

Furthermore, instead of starting with $x(t) = \cos(\omega t)$, let's start by finding the particular solution when the input is $x(t) = e^{\alpha t}$. The reason we want to do this will become apparent in a moment. As always, we get the particular solution by guessing an answer and then verifying that our guess satisfies the differential equation. Let's start by guessing that the system response will also be a decaying exponential with the same decay rate: $y(t) = Ke^{\alpha t}$. If that is the case, then $dy/dt = \alpha Ke^{\alpha t}$. Plugging into Equation [4-1] we find

$$\frac{dy}{dt} + \frac{y}{RC} = \frac{x}{RC}$$

$$\alpha Ke^{\alpha t} + \frac{K}{RC}e^{\alpha t} = \frac{1}{RC}e^{\alpha t}$$

$$\alpha K + \frac{K}{RC} = \frac{1}{RC}$$

$$K\left(\alpha + \frac{1}{RC}\right) = \frac{1}{RC}$$

$$K = \frac{1}{\alpha RC + 1} \qquad\qquad [4\text{-}8]$$

Therefore, if our input is $x(t) = e^{\alpha t}$, our steady-state output is

$$y(t) = \frac{1}{\alpha RC + 1} e^{\alpha t} \qquad [4\text{-}9]$$

We can now quickly solve for the particular solution in response to $x(t) = \cos(\omega t)$ by recalling that $\cos(\omega t) = 1/2 e^{j\omega t} + 1/2 e^{-j\omega t}$. Using Equation [4-9], we can see by inspection that the complete particular solution in response to the cosine input will be

$$y(t) = \frac{1}{2} \cdot \frac{1}{j\omega RC + 1} e^{j\omega t} + \frac{1}{2} \cdot \frac{1}{-j\omega RC + 1} e^{-j\omega t}$$

$$= A(\omega) \cos(\omega t + \theta(\omega)) \qquad [4\text{-}10]$$

where

$$A(\omega) = \left| \frac{1}{j\omega RC + 1} \right|$$

$$= \frac{1}{\sqrt{(\omega RC)^2 + 1}} \qquad [4\text{-}11]$$

and

$$\theta(\omega) = \angle \frac{1}{j\omega RC + 1}$$

$$= -\operatorname{atan}(\omega RC) \qquad [4\text{-}12]$$

So we have our answer. When the input to a filter is a cosine, *the output is also a cosine with the same frequency*. The amplitude and phase of the output signal vary as a function of frequency and are given by the Equations [4-11] and [4-12], respectively. A quick inspection of Equation [4-11] shows that as ω increases, $A(\omega)$ decreases. This is exactly what we expect for a low-pass filter! The amplitude of the output cosine decreases as the frequency of the cosine increases. This shows that we indeed have a low-pass filter.

4.2.2 Linear and Time Invariant Systems

This concept in which the system output is a cosine at the same frequency as the input is an incredibly powerful concept, and it is the underlying reason we spend so much time obsessing over cosines. Cosines are the *only* function with the property that the input and output signals have the same shape, differing only in the relative amplitudes and phase shifts. It turns out this will always be the case as long as the system in question is *linear* and *time-invariant*. A linear system has two properties. The first is that if you scale the input by some constant, then you scale the output by that same constant (i.e., doubling the input means doubling the output). The second is that if your input is the sum of two signals, then your output will be the sum of the system's individual responses to those two signals. Mathematically, a linear system can be expressed as

$$c_1 x_1(t) + c_2 x_2(t) \Rightarrow c_1 y_1(t) + c_2 y_2(t)$$ [4-13]

A time-invariant system is one in which the system's response does not depend on when the input is presented. In other words, delaying an input signal $x(t)$ by τ seconds simply means the output will be $y(t)$, also delayed by τ seconds, so

$$x(t - \tau) \Rightarrow y(t - \tau)$$ [4-14]

Any linear time-invariant (LTI) system will have the vitally important property that when the system's input is a cosine, the output will be a cosine at the same frequency, differing only in magnitude and phase.

Section 4.3 First-Order Systems—Bode Plots

So far, we've learned that Equation [4-11] describes how the magnitude of a low-pass filter varies as a function of frequency. This equation has all of the features we would expect for a low-pass filter. For example, its maximal value is $A = 1$ and occurs when $\omega = 0$. This means that at low frequencies ($\omega = 0$ rads/sec is as low as things go) the filter *gain* is one, meaning that the output magnitude is equal to one times the input magnitude. In other words, at low frequencies, signals are passed through the filter perfectly (more or less). Conversely, the minimum value of Equation [4-11] is $A = 0$, which occurs as $\omega \rightarrow \infty$. This means that at high frequencies (infinity is pretty high . . .), the filter rejects the signal; output magnitude equals zero times the input magnitude. In combination, these two traits confirm that we have a low-pass filter.

4.3.1 Linear Plot

Perhaps the most obvious way to proceed with our analysis of Equation [4-11] is simply to plot it (see Figure 4-5). The figure confirms that $A(\omega = 0) = 1$ and $A(\omega \rightarrow \infty) = 0$. This is most certainly a low-pass filter! It is interesting and absolutely critical to notice that, unlike the ideal filters shown in Figure 4-4, this filter does not have a crisp delineation between frequencies that are perfectly passed and frequencies that are absolutely rejected. Instead there is a sliding scale where many frequencies are reduced but not completely filtered away. For example at $\omega = 5$ rads/sec, the filter gain is about 0.2, meaning input signals at those frequencies are reduced to about 20% of their original amplitude but are not rejected completely. This is very typical of filters and we must always bear in mind that plenty of signal energy is passed even at frequencies beyond the cutoff frequency.

4.3.2 Log-Log Plot

Despite the fact that Figure 4-5 is technically accurate, it is highly unusual in practice to plot a filter's magnitude response in this way. Instead, it is far more common (and useful) to plot

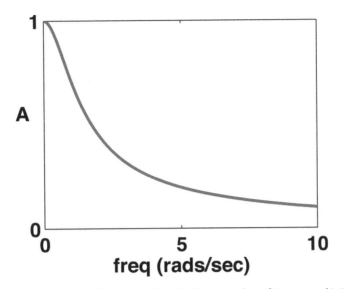

Figure 4-5: Magnitude plot of a low-pass filter with $RC = 1$. This is a plot of Equation [4-11].

Equation [4-11] as shown in Figure 4-6. This is called a *Bode Plot*. Note that both Figure 4-5 and Figure 4-6 are plots of Equation [4-11]. However, Figure 4-6 is a log-log plot, whereas Figure 4-5 is a linear-linear plot.

The x-axis of Figure 4-6 shows frequency plotted on a log scale. This tends to overemphasize low frequencies and underemphasize high frequencies; it stretches out the left side of the plot, which was not easily readable in Figure 4-5. The y-axis of Figure 4-6 plots the *decibels* of Equation [4-11]. Decibels are defined as

$$dB(A) = 20\log_{10}(A) \tag{4-15}$$

For example, we see that when $A = 1$, $dB(A) = 0$ dB. Zero decibels always means that "output equals input," or "zero change." At higher frequencies, the value of A decreases. For example,

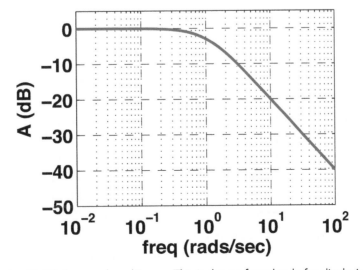

Figure 4-6: Plot of Equation [4-11] done on log-dB axes. This is the preferred style for displaying filter characteristics.

when $A = 0.5$, then $dB(0.5) = -6$ dB. Negative decibels imply that the output is smaller than the input. We also see that as A approaches zero, then $dB(A) \rightarrow -\infty$.

The plot in Figure 4-6 is the standard plot of the magnitude response of a filter. Note several important things. When ω is small, the plot of A is approximately constant at $A = 0$ dB. Conversely, when ω is large, the plot of A is a line with a negative slope. To understand where these lines come from, we need to take a closer look at Equation [4-11].

Equation [4-11] can be simplified rather nicely if we are willing to make some simple assumptions. For example, if we assume that ω is so small that $\omega RC \ll 1$, then we can ignore the first term in the denominator. The equation then simplifies to $A \approx 1 = 0$ dB. Conversely, if the frequency is so large that $\omega RC \gg 1$, then we can disregard the second term in the denominator and simplify to $A \approx 1/\omega RC$. Of course we prefer to give our filter gain in decibels, therefore $1/\omega RC$ can be expressed as

$$A \approx 20\log\left(\frac{1}{\omega RC}\right)$$

[4-16]

For the sake of argument, let's increase ω by a factor of ten (also known as one *decade*). Equation [4-16] now becomes

$$A \approx \frac{1}{10\omega RC}$$

[4-17]

Using the laws of logarithms, we can convert this to decibels as follows.

$$A \approx 20\log\left(\frac{1}{10\omega RC}\right)$$

$$= 20\log\left(\frac{1}{\omega RC}\right) - 20\log(10)$$

$$= 20\log\left(\frac{1}{\omega RC}\right) - 20dB$$

[4-18]

In comparing Equations [4-16] and [4-18], we see they differ by only the constant amount of 20 dB. Therefore, we have demonstrated that every time we increase the frequency by one decade, the filter response is reduced by 20 dB. On a log-dB scale (as is the case for Figure 4-6) this will plot as a line with slope −20 dB per decade. For example, in Figure 4-6, the y-value at $\omega = 10$ rads/sec is −20 dB. One decade later, at $\omega = 100$ rads/sec, the y-value is −40 dB. This −20 dB/dec slope is a very common and important value in signal processing engineering; you will see it time and again as your studies proceed.

Finally, we discuss the derivation of a cutoff frequency that delineates where the filter starts removing frequencies of signal. Looking at Figure 4-6, it's not immediately obvious where the cutoff frequency ought to go. Long ago, engineers decided that a good place to designate as the cutoff frequency is wherever the output signal has only half the power of the input signal. It turns out that, for a cosine, we reduce the power by half when we reduce the amplitude by a factor of $\sqrt{2}$. Therefore, since we know that at low frequencies the gain is $A = 1$, we can determine the cutoff frequency by taking Equation [4-11], setting $A = 1/\sqrt{2}$, and solving for ω.

$$A = \frac{1}{\sqrt{(\omega RC)^2 + 1}}$$

$$\frac{1}{\sqrt{2}} = \frac{1}{\sqrt{(\omega_c RC)^2 + 1}}$$

$$2 = (\omega_c RC)^2 + 1$$

$$\omega_c = \frac{1}{RC} \tag{4-19}$$

where the subscript "c" identifies ω_c as the cutoff frequency. Equation [4-19] is a superb finding, as it validates what we were predicting at the end of the last section. Specifically, we predicted that the cutoff frequency would have to decrease as RC increased, and vice versa. Equation [4-19] shows that this is clearly the case.

Another interesting property of the cutoff frequency is that it occurs when the output signal magnitude is $1/\sqrt{2}$ or about 0.707 as big as the input signal. If we convert this to decibels, we get approximately −3 dB. Just like the −20 dB/decade number, −3 dB is one of those "magic" numbers that keep reappearing in signal processing. In fact, the cutoff frequency is often simply referred to as the −3 dB point.

Section 4.4 First-Order Systems—Square Wave and Harmonics

Let's try using our RC low-pass filter system to filter the square wave shown in Figure 4-7. Note that the period is $T = 1$ second.

For the filter, we select (somewhat arbitrarily) a cutoff frequency of $\omega_c = 6\pi$ rads/sec, meaning that $RC = 1/\omega_c = 1/6\pi$ seconds. Before we start, we should acknowledge that we pretty much already know what the filter output $y(t)$ is going to look like, because we've already done a hand calculation for the filter step response, which we found to be $y(t) = 1 - e^{-t/RC}$. Therefore, we are

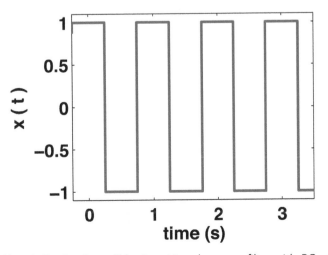

Figure 4-7: Square wave $x(t)$ with $T = 1$ s that will be input to a low-pass filter with $RC = 1/6\pi$ s.

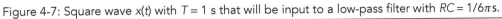

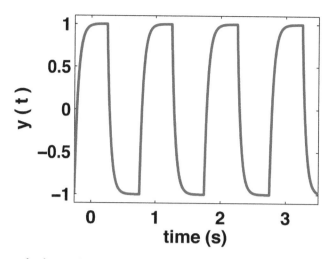

Figure 4-8: Expected response of a first-order low-pass filter with $RC = 1/6\pi s$ to the input square wave shown in Figure 4-7.

reasonably confident that when we are done we should arrive at the output signal shown in Figure 4-8.

Our analysis will proceed in three steps:

1.   Use Fourier analysis to express $x(t)$ as an infinite sum of cosines.
2.   Use filter theory to low-pass filter each of the cosine signals comprising $x(t)$.
3.   Sum the filtered cosines to produce the predicted filter output $y(t)$.

Fortunately, we learned to solve the first step when we studied Fourier Series Chapter 2. Without repeating all the math, we can express the Fourier Series of $x(t)$ as

$$A_n = \frac{\sin(\pi n/2)}{\pi n/2} \qquad \text{[4-20]}$$

where

$$\omega = \frac{2\pi n}{T} = 2\pi n \qquad \text{[4-21]}$$

This means that the signal has energy at the integer harmonics of $\omega = 2\pi$ rads/sec. The Fourier coefficients of the first few harmonics are summarized in Table 4-1.

Table 4-1: Fourier coefficients of the square wave in Figure 4-7

n	ω	A_n
1	2π	$\frac{2}{\pi} = \frac{2}{\pi}e^{j0}$
2	4π	0
3	6π	$-\frac{2}{3\pi} = \frac{2}{3\pi}e^{j\pi}$
4	8π	0
5	10π	$\frac{2}{5\pi} = \frac{2}{5\pi}e^{j0}$

Using Table 4-1, we can express the first few terms of $x(t)$ as the sum

$$x(t) \approx \frac{4}{\pi}\cos(2\pi t) + \frac{4}{3\pi}\cos(6\pi t + \pi) + \frac{4}{5\pi}\cos(10\pi t) + \dots \qquad [4\text{-}22]$$

Therefore, we have expressed our input signal $x(t)$ as an infinite sum of cosines. Now we will proceed with the task of filtering each of those cosines. Recall that when a cosine is input into a low-pass filter, the output signal is a cosine of the same frequency, whose magnitude is multiplied by $1/\sqrt{(\omega RC)^2 + 1}$ with a phase shift of $-atan(\omega RC)$. Remember that for this example, $RC = 1/6\pi$. For the first term in Equation [4-22], the cosine's frequency is $\omega = 2\pi$. Therefore we expect the filter response to be $A = 1/\sqrt{\left(\frac{2\pi}{6\pi}\right)^2 + 1} = 0.95$ and $\theta = -atan(2\pi/6\pi) = -0.32$ rads. Therefore, the filtered version of the first term of Equation [4-22] will be

$$\frac{4}{\pi} \cdot 0.95\cos(2\pi t - 0.32)$$

In a similar vein, we can filter the other two terms of Equation [4-22] and arrive at

$$y(t) \approx \frac{4}{\pi} \cdot 0.95\cos(2\pi t - 0.32) + \frac{4}{3\pi} \cdot 0.71\cos(6\pi t + \pi - 0.78) + \frac{4}{5\pi} \cdot 0.5\cos(10\pi t - 1) + \dots \ [4\text{-}23]$$

(see if you can calculate for yourself the values of 0.71 and −0.78 for $\omega = 6\pi$, and 0.5 and −1 for $\omega = 10\pi$).

Filter theory predicts that if you individually filter each of the infinite cosines comprising $x(t)$ using an LTI filter and then sum the resulting filtered cosines, you should get back the complete filtered signal $y(t)$ as predicted in Figure 4-8. Figure 4-9 plots the function of Equation [4-23], proving that the filtered cosines sum to the predicted step response.

Amazingly, the figure includes only the first three harmonics (as shown in Equation [4-23]) and already the signal is clearly the predicted step response. Upon reflection, this shouldn't be so surprising—by filtering away the high-frequency signal components, we need only the very lowest harmonics to reconstruct our step response. Our filter cutoff frequency is $\omega_c = 1/RC = 6\pi$

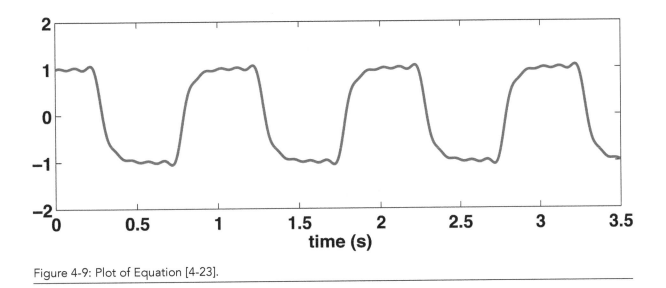

Figure 4-9: Plot of Equation [4-23].

rads/sec. Therefore, there won't be much energy left in the signal $y(t)$ past, say, the 5th harmonic at $\omega = 10\pi$ rads/sec.

Section 4.5 Fourier Analysis of Differential Equations

Until now, our efforts to predict filter responses have depended on our ability to find solutions to the system differential equation (Equation [4-1]). That process relied on some unsavory calculus. Let's see if there's an easier approach.

4.5.1 Differential Equations

Suppose we want to take the Fourier Transform of Equation [4-1]. Let's assume that the transform of $x(t)$ is $X(j\omega)$ and the transform of $y(t)$ is $Y(j\omega)$. You might recall that taking the derivative of a function is equivalent to multiplying its Fourier Transform by $j\omega$. Therefore the transform of dy/dt will be $j\omega Y(j\omega)$. Putting all this together, the Fourier Transform of Equation [4-1] will be

$$j\omega Y(j\omega) + \frac{Y(j\omega)}{RC} = \frac{X(j\omega)}{RC} \tag{4-24}$$

We can now combine like terms.

$$Y(j\omega)\left[j\omega + \frac{1}{RC} \right] = \frac{X(j\omega)}{RC}$$

$$Y(j\omega)[j\omega RC + 1] = X(j\omega)$$

$$Y(j\omega) = \frac{1}{j\omega RC + 1} X(j\omega) \tag{4-25}$$

By convention, we define the *transfer function* for our low-pass filter to be

$$H(j\omega) = \frac{Y(j\omega)}{X(j\omega)} = \frac{1}{j\omega RC + 1} \tag{4-26}$$

which also means that

$$Y(j\omega) = H(j\omega)X(j\omega) \tag{4-27}$$

Using our rules for computing the magnitude and phase of a complex expression, we find

$$|H(j\omega)| = \frac{1}{\sqrt{(\omega RC)^2 + 1}} \tag{4-28}$$

and

$$\angle H(j\omega) = -\mathrm{atan}(\omega RC) \qquad [4\text{-}29]$$

Happily, these are exactly the formulas we derived earlier after a lot of hard calculus for solving the differential equation! Hopefully you will agree that this method is far less painful.

4.5.2 System Functionality—Frequency

Equation [4-27] merits some more discussion. You will notice that in the frequency space, the output signal $Y(j\omega)$ is the product of the input signal $X(j\omega)$ and the transfer function $H(j\omega)$. All three of these functions are *complex*. Recall that when you multiply two complex expressions, you multiply the magnitudes and add the phases. For example, if $a = 4e^{j\pi/3}$ and $b = 3e^{j\pi/4}$, then $ab = 12\, e^{j7\pi/12}$. What Equation [4-27] tells us is that if our input is a cosine, then our output is also a cosine (at the same frequency, ω) whose magnitude has been *multiplied* by $|H(j\omega)|$, and whose phase has been *added* to $\angle H(j\omega)$. This is exactly what we did in Section 4.2.1.

The technique presented in this section can be used to easily derive the transfer function of *any* filter, provided we know its differential equation. In the next section, we'll advance this technique one step further and eliminate the need to even derive the differential equation in the first place.

Figure 4-10 summarizes what we've learned about signal processing so far. Our goal thus far has been to predict the output signal $y(t)$ that arises when input signal $x(t)$ is passed through a system. Figure 4-10 indicates that one way to do this is to take the Fourier Transform of the signal $x(t)$ and multiply it by the system's Transfer Function $H(j\omega)$ to get the Fourier Transform of the output signal, or $Y(j\omega)$. Finally, we take the inverse Fourier Transform of $Y(j\omega)$ to arrive at the output signal $y(t)$. Interestingly, Figure 4-10 appears to indicate that there may be a procedure for calculating $y(t)$ without even using the frequency domain in the first place. In other words, there should be some mathematical function for combining $x(t)$ and $h(t)$ (which we'll address in Section 4.6) to produce $y(t)$ directly. That function is called convolution and will be the subject of Chapter 5.

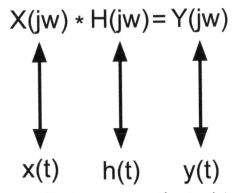

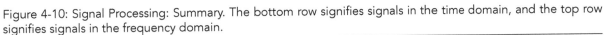

Figure 4-10: Signal Processing: Summary. The bottom row signifies signals in the time domain, and the top row signifies signals in the frequency domain.

4.5.3 Complex Impedances

You might recall that Ohm's Law relates voltage and current in a resistor.

$$v(t) = i(t)R \qquad \text{[4-30]}$$

If we take the Fourier Transform of both sides of the Ohm's Law equation, we arrive at

$$V(j\omega) = I(j\omega)R \qquad \text{[4-31]}$$

We can think of this as Ohm's Law in the frequency domain. Next we consider a capacitor. Every electrical engineering student should know the voltage-current relationship for a capacitor, as given by Equation [4-32].

$$i(t) = C\frac{dv(t)}{dt} \qquad \text{[4-32]}$$

Again, we take the Fourier Transform of this equation, recalling that in the transform-space, the derivative function becomes a simple multiplication by $j\omega$.

$$I(j\omega) = Cj\omega V(j\omega)$$

$$V(j\omega) = I(j\omega)\frac{1}{j\omega C} \qquad \text{[4-33]}$$

Comparing Equations [4-31] and [4-33], we see there is a nice parallel. Both appear to be versions of Ohm's Law in the frequency domain, where $V(j\omega) = I(j\omega) Z(j\omega)$. The term $Z(j\omega)$ (or just Z) is the *complex impedance* of the circuit element. The impedance of a resistor is R, and the impedance of a capacitor is $1/j\omega C$.

Now that everything can be interpreted as an application of Ohm's Law, circuits become much easier to solve. Recalling the simple first-order low-pass filter shown in Figure 4-1, we can now solve for the output using Ohm's Law, complex impedances, and Kirchoff's Current Law.

$$\frac{X(j\omega) - Y(j\omega)}{R} = \frac{Y(j\omega)}{1/j\omega C}$$

$$X(j\omega) - Y(j\omega) = j\omega RC Y(j\omega)$$

$$X(j\omega) = Y(j\omega)\left[1 + j\omega RC\right]$$

$$Y(j\omega) = \frac{1}{1 + j\omega RC} X(j\omega) \qquad \text{[4-34]}$$

We can rewrite Equation [4-34] in the more general form of

$$Y(j\omega) = H(j\omega) X(j\omega) \qquad \text{[4-35]}$$

where, for the case of the first-order low-pass filter in Figure 4-1,

$$H(j\omega) = \frac{1}{j\omega RC + 1}$$ [4-36]

This is terrific! We are able to solve for the Transfer Function with no more than a couple lines of algebra. This is quite an improvement over solving the differential equation the hard way.

Section 4.6 Impulse Response

One interesting property of a circuit is known as its *impulse response*. As the name suggests, the impulse response is the circuit's output when the input is an instantaneous impulse, or $x(t) = \delta(t)$. Impulse responses are interesting to study because, as we learned in Section 3.3.2, impulses contain equal energy at all frequencies. Therefore, studying the impulse response allows us to study the system's response to all frequencies in much the same way a Bode plot does. By convention, we always denote the impulse response as $h(t)$.

You might recall that the derivative of the step function is the impulse function $\delta(t)$. We can use this property to quickly solve for the impulse response: if the derivative of the step function is the impulse function, then the derivative of the step response is the impulse response.

4.6.1 First-Order Systems—Impulse Response

You might recall that the step response for our low-pass filter circuit is $y(t) = 1 - e^{-t/RC}$. Therefore, the impulse response is

$$h(t) = \frac{d}{dt}(\text{step response})$$

$$= \frac{d}{dt}\left(1 - e^{-t/RC}\right)$$

$$= \frac{1}{RC} e^{-t/RC}$$ [4-37]

4.6.2 Fourier Transform of Impulse Response

Next, just for fun, let's take the Fourier Transform of the impulse response. Note that since the impulse is presented at $t = 0$, there is no output signal for $t < 0$ and therefore nothing to integrate.

$$FT\{h(t)\} = \int_0^\infty \frac{1}{RC} e^{-t/RC} e^{-j\omega t} \, dt$$

$$= \frac{1}{RC} \int_0^\infty e^{-t\left(j\omega + \frac{1}{RC}\right)} \, dt$$

$$= \frac{1}{RC} \int_0^\infty e^{-\frac{j\omega RC + 1}{RC} t} \, dt$$

$$= \frac{1}{RC} \cdot \frac{-RC}{j\omega RC + 1} \left[e^{-\frac{j\omega RC + 1}{RC} t} \right]_0^\infty$$

$$= \frac{-1}{j\omega RC + 1}[0-1]$$

$$= \frac{1}{j\omega RC + 1} \hspace{3cm} [4\text{-}38]$$

You might recognize Equation [4-38] as the Transfer Function of our circuit. This is an amazing finding! It appears that the Fourier Transform of the impulse response is the Transfer Function. It turns out that this will *always* be true, regardless of what circuit we are analyzing.

Upon reflection, this result actually makes intuitive sense. As stated earlier, the impulse contains equal energy at all frequencies. Then, when the impulse is passed through a filter, some frequencies will remain while others are filtered out; the impulse response h(t) should therefore clearly indicate which frequencies were passed and which were rejected. No wonder then that when we examine the frequency content of h(t) by taking its Fourier Transform, we get the transfer function $H(j\omega)$, which, by definition, tells us which frequencies are passed and which are rejected. This relationship will always be the case for any LTI system.

Section 4.7 Matlab Control Systems Toolbox

So far we've spent a lot of time learning how to do hand calculations to describe filters and to calculate filter responses. The good news is that Matlab has a number of tools for describing and applying filters. These tools are mostly in the Control Systems Toolbox. This section will show you some of the most useful features and how to use them.

4.7.1 Defining the Filter

Suppose our goal is to analyze a low-pass filter circuit like the one shown Figure 4-1. We know from experience that the transfer function is given by $H(j\omega) = 1/(j\omega RC + 1)$. In Matlab, we can define this filter as follows (using sample values of $R = 1k\Omega$ and $C = 1\mu F$)

```
R = 1000;
C = 1e-6;
H = tf ( 1 , [R*C 1]);
```

The tf command creates a *transfer function* using 1 as the numerator and $j\omega$RC + 1 for the denominator. Note that, as far as the tf function is concerned, the $j\omega$ are assumed and don't need to be explicitly placed in the definition of H.

4.7.2 Plotting the Transfer Function

Now that we've defined H, we can do a number of really useful things. For example, we can have Matlab create the transfer function plots (magnitude and phase) using the bode function.

```
bode (H);
```

This creates the plot seen in Figure 4-11.

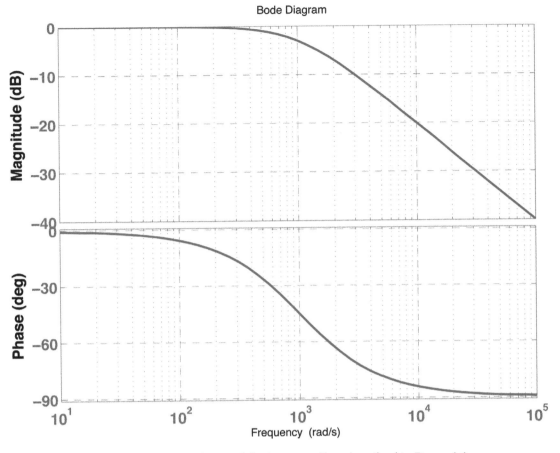

Figure 4-11: Bode plot (magnitude and phase) of the low-pass filter described in Figure 4-1.

As useful as this is, perhaps we are interested only in finding the magnitude and phase at a handful of specific frequencies. In that case, we can use the bode function with an explicit list of frequencies.

```
w = [100 1000 10000];

[M,P] = bode(H,w);
disp(M(:));
disp(P(:));
```

The results returned from Matlab are:

```
0.9950
0.7071
0.0995

-5.7106
-45.0000
-84.2894
```

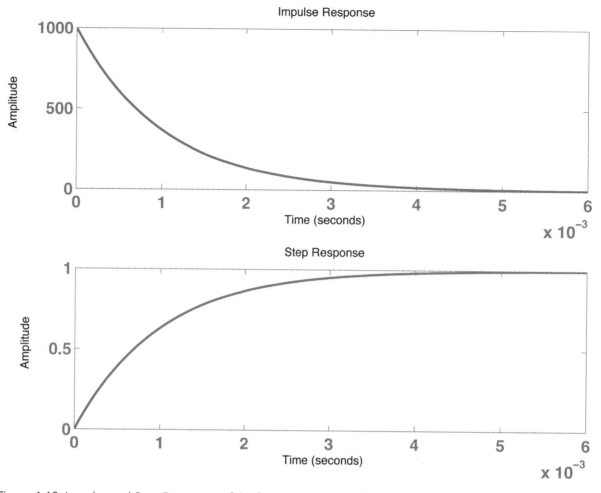

Figure 4-12: Impulse and Step Responses of the first-order low-pass filter with $R = 1$ KΩ and $C = 1$ μF.

Therefore, we see that the Magnitude of the filter response for $\omega = 10$ rads/sec is 0.995 (about −0.0432 dB), and the phase at that frequency is −5.7106 degrees. We also have magnitude and phase outputs for 1,000 and 10,000 rads/sec. You can (and should) validate that these are the correct answers (using $M = 1/\sqrt{(\omega RC)^2 + 1}$ and $= -\text{atan}(\omega RC)$). You should also verify that these results are consistent with the plots shown in Figure 4-11.

4.7.3 Impulse and Step Response

You can also use Matlab to automatically plot the impulse and step responses of a filter. This can save you loads of tedious hand calculations involving especially unseemly calculus.

- `impulse(H);` creates the impulse response
- `step(H);` creates the step response

Running either of these commands automatically creates the respective plots; these are shown in Figure 4-12. We are pleased to see the plots we expect—the step response is clearly $1 - e^{-t/RC}$ and the impulse response is $\frac{1}{RC}e^{-t/RC}$.

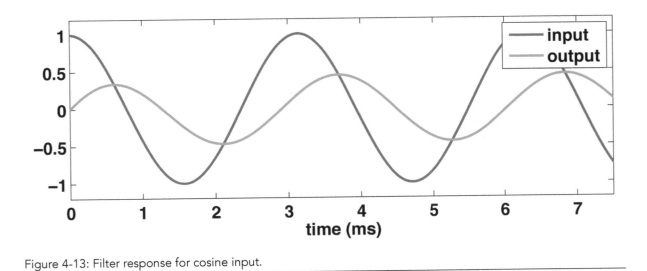

Figure 4-13: Filter response for cosine input.

4.7.4 Simulating the Filter

Perhaps the most interesting thing you can do with the control system toolbox is to quickly and easily simulate what the filter output would be in response to an arbitrary input. To do this, you need the lsim command. The first step is to create the input signal. This requires defining both a time vector, which we'll call t, and an input signal vector, which we'll call x. You can define any input vector you want. Here are a few examples.

```
x = cos(2000*t);        % cosine at w=2000rads/s
x = square(4000*t);     % square wave at w=4000rads/s
x = randn(size(t));     % random noise
```

Once you have declared x and t, it's time to simulate the filter response using the lsim command. Fortunately, it's pretty easy.

```
y = lsim(H,x,t);
```

Finally, plot your answers and make sure the results make sense.

```
plot(t,x,t,y);
xlabel('time');
legend('input','output');
```

Note that all of the Matlab functions shown in this section can be used in a variety of ways depending on which inputs and outputs are specified. As always, use the Matlab help function to find out more. For example, type help impulse to see how many ways there are of using the impulse function.

Section 4.8 Pole-Zero Theory

4.8.1 Laplace Transform

So far, we've made excellent use of the Fourier Transform to simplify the math for interpreting systems. In this section, we'll introduce a related transform, called the *Laplace Transform* that is even more helpful in filter design and analysis.

As you might recall, the Fourier Transform is defined as

$$F(j\omega) = \int x(t)e^{-j\omega t}dt$$

The point of the Fourier Transform is to express signal $x(t)$ as the weighted sum of complex exponentials. The frequency of the exponential terms are all *purely imaginary*: i.e., the frequencies are all of the form $j\omega$. The Laplace Transform works almost exactly like the Fourier Transform, except that it allows for the frequencies of the exponentials to be *complex* as opposed to just purely imaginary. In other words, instead of limiting ourselves to frequencies of $j\omega$, the Laplace Transform allows us to have frequencies of $(\sigma + j\omega)$, which has both a real (σ) and imaginary ($j\omega$) part.

It turns out that writing out $\sigma + j\omega$ gets old pretty quickly, so we introduce the shorthand of $s = \sigma + j\omega$ instead. Using this abbreviation, the formal definition of the Laplace transform is

$$L(s) = \int x(t)e^{-st}dt \qquad\qquad [4\text{-}39]$$

You will be pleased to learn that all the intuition and properties of the Fourier Transform carry over pretty much identically to the Laplace Transform. You will also note that by setting $\sigma = 0$, the Laplace Transform collapses to the Fourier Transform: the latter is a subset of the former.

To see how the Laplace Transform can be used to study a filter, let's consider the case of the first-order low-pass filter we've been studying. You might recall that the impulse response of this filter was given by the expression $h(t) = (1/RC)e^{-t/RC}$. You might also recall that taking the Fourier Transform of the impulse response leads to the Transfer Function. In this example, we'll take the Laplace Transform of the impulse response and explore the result:

$$H(s) = \int x(t)e^{-st}dt$$

$$= \int_0^\infty \frac{1}{RC}e^{-t/RC}e^{-st}\,dt$$

$$= \frac{1}{RC}\int_0^\infty e^{-t\left(s+\frac{1}{RC}\right)}\,dt$$

$$= \frac{1}{RC}\cdot\frac{-1}{s+\dfrac{1}{RC}}\left[e^{-t\left(s+\frac{1}{RC}\right)}\right]_0^\infty$$

$$= \frac{1}{RC} \cdot \frac{-1}{s + \frac{1}{RC}} [0-1]$$

$$= \frac{1/RC}{s+1/RC} \qquad \text{[4-40]}$$

Recalling that for our first-order low-pass filter, the cutoff frequency is $\omega_c = 1/RC$, we can simplify and complete our calculation.

$$H(s) = \frac{\omega_c}{s + \omega_c} \qquad \text{[4-41]}$$

Transfer functions in the form of Equation [4-41] are very common in signal processing and it's worth learning how to dissect them. The obvious thing we might want to do is to calculate the magnitude response of the filter. We do this by taking the Fourier Transform of our filter, which is achieved by setting $\sigma = 0$, and taking the magnitude.

$$\left| H(j\omega) \right| = \frac{\omega_c}{\sqrt{\omega^2 + \omega_c^2}} \qquad \text{[4-42]}$$

In theory, we could use a computer to plot this function, but that would be far too much work and it yields little intuition. A better method is described here.

4.8.2 Poles and Zeros

We start by solving for the *poles* and *zeros* of the filter. Poles are values of s that make the denominator of Equation [4-41] equal to zero (i.e., the roots of the denominator), whereas zeros

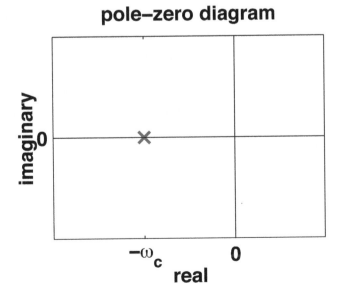

Figure 4-14: Pole-zero diagram corresponding to the filter of Equation [4-41]. There is only a single pole in this case.

are values of s that make the numerator equal to zero (i.e., the roots of the numerator). In the case of Equation [4-41], there are no zeros, and just one pole, located at $s = -\omega_c$.

A common procedure is to create a *pole-zero diagram*. This is a diagram of the complex s-plane and it shows the relative locations of the poles and zeros. By convention, poles are always denoted with an x and zeros with an o. The pole-zero diagram of Equation [4-41] is shown in Figure 4-14.

Note that the pole has been located on the real axis, where $s = -\omega_c$. Now, recall that cosines, which we use to construct all real-valued functions, consist only of purely imaginary frequencies. In other words, if $s = \sigma + j\omega$, cosines occur only when $\sigma = 0$, which for a pole-zero diagram will always correspond to the y-axis. Therefore, if we want to know the system response to a cosine at, say $\omega = 100$ rads/sec, we put our pencil on the y-axis location corresponding to $j\omega = 100$ and we see how far that point is from the poles and zeros. If that point is close to a zero, we expect that the system response will be close to zero at that frequency. If the point is close to a pole, we expect the magnitude of the system response will be driven up away from zero. The value of the pole-zero diagram is that it provides for a quick and intuitive assessment of the transfer function. For example in Figure 4-14, we see that at low frequencies, the magnitude of the transfer function is propped up by the pole. But then as we travel up the $j\omega$ axis, we move farther from the pole and the magnitude response drops off. This is consistent with our understanding of a low-pass filter's behavior.

In interpreting the pole-zero diagram, it can be helpful to imagine a circus-tent top being draped over the complex s-plane. Everywhere there is a pole, you can imagine an actual pole propping up the tent. Everywhere there is a zero, you can imagine the tent being staked into the ground. To estimate the transfer function, you then start at the origin and take a walk up the $j\omega$ axis; the height of the tent over your head will correspond to the magnitude of the transfer function as a function of frequency.

There is no limitation on the number of poles and zeros a filter can have. So far, we've looked at only single-pole filters, but you can easily have more poles. Using a filter with two poles is roughly equivalent to filtering a signal twice: you can expect better filtering. Generally speaking, more poles and zeros translates to more precise filtering.

Poles and zeros are also very handy because they provide us with a nice shortcut for creating the Bode plot by hand without having to do any sophisticated math. The process is simple: every time you encounter a pole, you decrease the slope of the Bode plot by 20dB/decade. Every time you encounter a zero, you increase the slope by 20dB/decade. In the case of the low-pass RC circuit, we start with a gain of 0dB and remain flat until we reach the cutoff frequency ω_c. Once we reach ω_c the slope decreases from being flat (0dB/decade) to −20dB/decade. Since this is just an estimate, we'll have to make some slight adjustments, but overall the process is relatively painless and yields a good degree of intuition.

Let's try a more specific example. Suppose

$$H(s) = \frac{10}{s^2 + 11s + 10} \tag{4-43}$$

We start by factoring the denominator, which in Matlab can be achieved by typing `roots([1 11 10])`.

$$H(s) = \frac{10}{(s+1)(s+10)} \tag{4-44}$$

pole–zero diagram

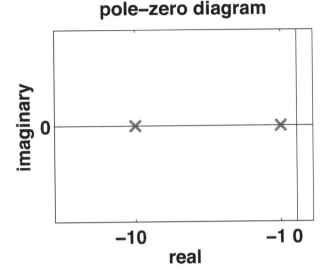

Figure 4-15: Pole-zero diagram corresponding to the filter of Equation [4-44]. There are two poles in this case.

Therefore we have no zeros and two poles at $s = -1, -10$ rads/sec. The pole-zero diagram is shown in Figure 4-15.

Now we use the poles to guide the quick-and-dirty method of creating magnitude plot. Following the steps above, we can quickly arrive at Figure 4-16. As Figure 4-16 demonstrates, we start by drawing a horizontal line. Then at $\omega = 1$ rads/sec, we reduce the slope of the line by 20 dB/decade, and then at $\omega = 10$ rads/sec, we reduce the slope again by another 20 dB/decade. The only thing left to do is to plug some numbers into Equation [4-44] in order to figure out how to label the y-axis. Let's try to guess what the magnitude response will be like for very low frequencies (i.e., where Figure 4-16 is flat). Look at Equation [4-44] and think about what this equation will do when ω is *really* small. In that case, we can simply reduce $s+1$ to just "1" (because $s = \sigma + j\omega$; we set σ to zero because we are solving for the transfer function and ω to

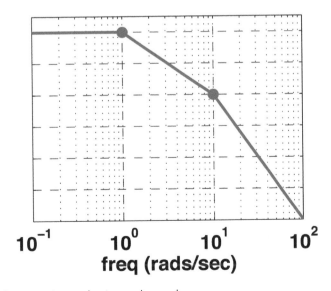

Figure 4-16: Magnitude plot as estimated using poles and zeros.

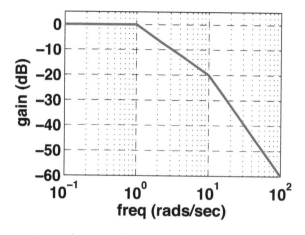

Figure 4-17: Magnitude plot as estimated using poles and zeros, with y-axis values filled in.

zero because we are investigating low frequencies). Likewise, we can reduce $s+10$ to just "10" for the same reason. Therefore, when ω is small, Equation [4-44] simplifies to $H(s) = 10/10 = 1$. Recalling that a gain of 1 corresponds to 0 dB, we can add values to the y-axis in order to complete our plot, shown in Figure 4-17.

Note that in Figure 4-17, the gain drops from 0 to -20 dB from $\omega = 1$ to $\omega = 10$, which is exactly a slope of -20 dB/decade. Likewise, the gain drops from -20 to -60 dB going from $\omega = 10$ to $\omega = 100$, which is a slope of -40 dB/decade, exactly as predicted.

The value in this approach is that we are very quickly able to sketch the magnitude response without having to get bogged down in the details of using a calculator or a computer. In fact, our plot is pretty accurate. Figure 4-18 superimposes our estimated transfer function plot from Figure 4-17, with the "true" correct plot, as generated by Matlab.

4.8.3 Complex Poles

Let's continue our analysis of poles and zeros but we introduce a new wrinkle: complex poles. Consider the transfer function

$$H(s) = \frac{2}{s^2 + 2s + 2} \qquad [4\text{-}45]$$

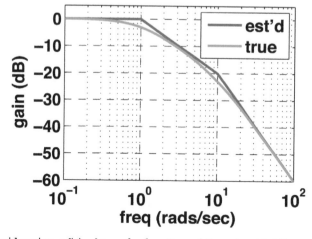

Figure 4-18: Estimated [black] and true [blue] transfer function of Equation [4-44].

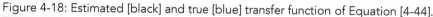

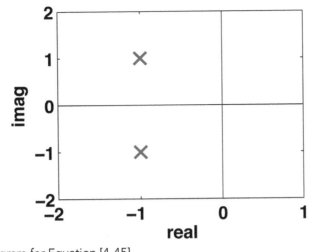

Figure 4-19: Pole-Zero diagram for Equation [4-45].

If we factor the denominator, we find that we have poles at $s_p = -1 \pm j = \sqrt{2} \cdot e^{\pm j3\pi/4}$. Note that the poles are complex conjugates; this will *always* be the case for complex poles. The pole-zero diagram is shown in Figure 4-19.

Unfortunately, our intuition for building the Bode plot isn't quite as robust as it was back when the poles were simply real. Fortunately, we still can use Matlab to quickly and easily tell us what kind of filter we have.

```
H = tf(2,[1 2 2]);
bode(H);
```

The resulting plot is shown in Figure 4-20; we appear to have a low-pass filter that cuts off around $\omega = \sqrt{2}$ rads/sec. Note that because we have two poles, the filter rolls off at −40 dB/decade. In general, for complex poles, we can expect the filter roll-off to start at approximately the magnitude of the pole frequency ($\omega = \sqrt{2}$ rads/sec in this case).

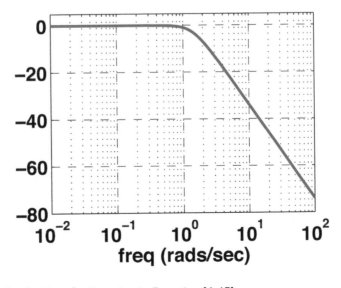

Figure 4-20: Bode Plot for the Transfer Function in Equation [4-45].

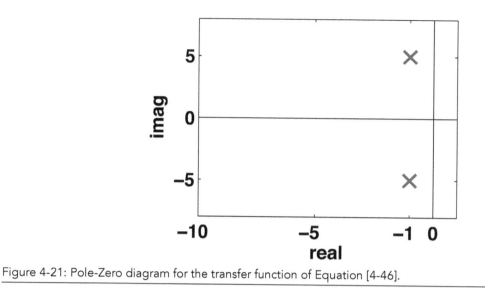

Figure 4-21: Pole-Zero diagram for the transfer function of Equation [4-46].

If we move these poles closer to the imaginary axis, something very peculiar occurs. Consider the transfer function

$$H(s) = \frac{26}{s^2 + 2s + 26}$$

[4-46]

We see that there are no zeros, and a pair of complex conjugate poles at $s_p = -1 \pm 5j$. The pole-zero diagram is shown in Figure 4-21. With the poles much closer to the imaginary $j\omega$ axis, our intuition tells us to expect the poles to have a strong impact on the transfer function, especially around $\omega = 5$ rads/sec. A plot of the magnitude response (i.e., the magnitude of Equation [4-46]) shows us what we have anticipated. Note that at high frequencies, the filter rolls off at -40dB/decade, in accordance with the fact that there are two poles. Complex poles are incredibly useful because they allow us to control the shape of the transfer function with great precision. In the next section, we'll learn how sets of complex poles can be combined to produce sophisticated filters with different properties.

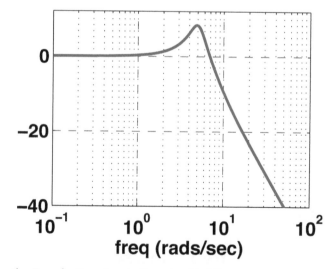

Figure 4-22: Bode Plot for the Transfer Function in Equation [4-46].

A final word about poles. You might have noticed that all our poles were in the left-hand s-plane. This is not an accident. All *stable* filters are characterized by left-hand plane poles. Consider a complex exponential signal e^{st} where $s = \sigma + j\omega$. A simple substitution and simplification tells us that an equivalent representation is $e^{\sigma t}e^{j\omega t}$. There are two parts to this product. The $e^{j\omega t}$ term, as we've learned, yields a cosine at frequency ω. The $e^{\sigma t}$ term, however, is a function that will either decay to zero (if $\sigma < 0$) or blow up to infinity (if $\sigma > 0$). Any function that blows up to infinity is, by definition, unstable. That means that if a system had a right-hand plane pole (i.e., $\sigma > 0$), we would expect that any input would destabilize the system and eventually result in an infinite output. In contrast, when you perturb a system with left-hand plane poles, the effects of the perturbation eventually die out or reach steady-state. This is a desirable property and is the very definition of a stable system.

Section 4.9 Introduction to Filters

While we are free to put poles anywhere in the left-hand s-plane that we like, it turns out that there are three special families of filters that are commonly used because they have some nice properties that are convenient for signal processing. These are called Butterworth, Chebyshev,

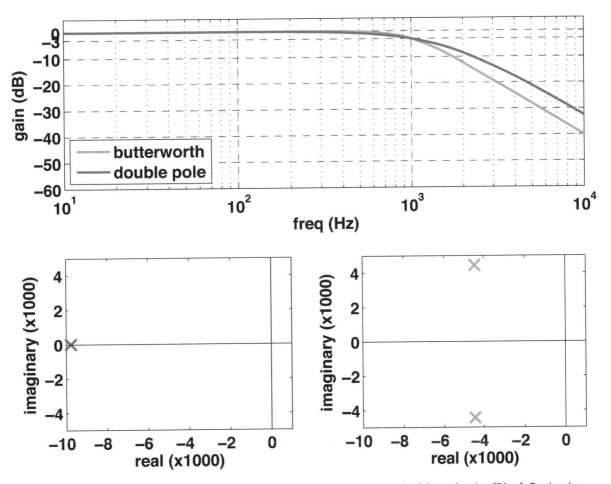

Figure 4-23: Comparison of two 2-pole filters. Butterworth [Green] versus double real poles [Blue]. Bode plots are shown at top, and pole-zero diagrams are shown at bottom.

and Elliptic filters. The details of their theory are beyond the scope of this book, but we will learn how to design them and what are the pros and cons of each one.

Let's start by looking at a basic Butterworth filter. This example will show us why it's not always desirable to go with just real poles; complex poles can be very useful if situated properly. Figure 4-23 shows two filters. Both filters have two poles, and both have the same cutoff frequency $f_c = 1000$ Hz. The "double pole" filter has two poles at exactly the same frequency (not that it's important, but the poles are both located at $s = -9762$). In comparison, the Butterworth filter has its poles at $s = -4443 \pm j4443$. Figure 4-23 shows the two Bode plots. The result is impressive: both filters have a cutoff frequency of exactly 1000 Hz, and both roll off at -40 dB/decade (you can tell it's the same roll-off because the two lines are perfectly parallel at high frequencies). However, the Butterworth filter is definitely doing a better job at rejecting high-frequency signals. For example, at f = 3000 Hz, the double-pole filter has a gain of about -13dB = 0.2, whereas the Butterworth filter has a gain of -19 dB = 0.11. The Butterworth filter is therefore doing a better job at rejecting all frequencies above the filter cutoff. There definitely appears to be some value in using complex poles. Let's explore further.

Figure 4-24 compares filter performance for Butterworth, Chebyshev, and Elliptic filters. As in the previous example, all three filters were designed to have two poles and a cutoff

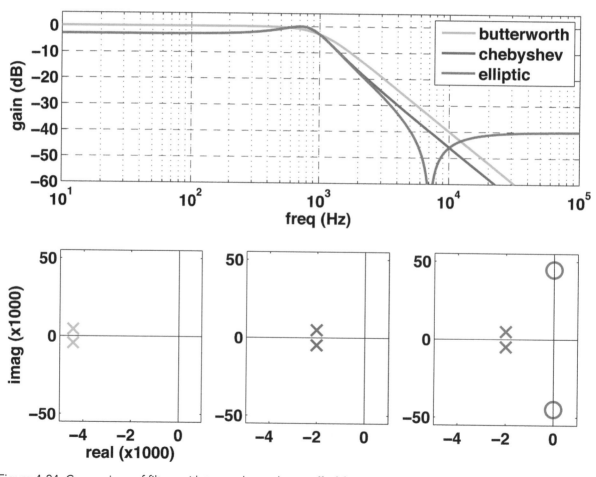

Figure 4-24: Comparison of filters with two poles and a cutoff of $f_c = 1000$ Hz.

frequency of 1000 Hz. The pole-zero diagrams clearly show that the poles have been placed in different locations for the three filters. The Bode plot shows some interesting behavior. Starting with the Butterworth filter [cyan], we can see that is very flat in the pass band. However, in the stop band (i.e., when $f > 1000$ Hz), the Butterworth filter is the least effective at rejecting signals.

In contrast, consider the Chebyshev filter [blue] has better performance in the stop band versus Butterworth. However, that improved performance came at a cost, specifically, the cost of some "passband ripple." As you can see, the Chebyshev filter isn't very flat in the passband; the filter response wiggles up and down, in this case between 0 and −3 dB.

Finally, the Elliptic filter [red], has not only the same passband ripple as Chebyshev, but also has some new behavior in the stop band. Specifically, we design the filter by specifying an acceptable amount of rejection in the stop band (−40 dB in this case). You can see that at frequencies greater than about 10000 Hz, the Elliptic filter is outperformed by Butterworth and Chebyshev, but it always satisfies its design criteria of having a stop band of at least −40 dB. In exchange for some flexibility in the stop band, the Elliptic filter outperforms Butterworth and Chebyshev in the *transition band*, meaning $1000 < f < 10000$ Hz. The transition band is where the filter response transitions between the passband and the stop band. The Elliptic filter achieves its performance by the inclusion of a pair of zeros, which act to reduce filter gain in the transition band.

So which filter is better? It depends on your particular application. The Chebyshev and Elliptic filters are more selective and can do a better job of rejecting high frequencies, but you have to be able to live with some passband ripple. Passband ripple means that there would be some distortion in the part of the signal you are trying to keep. Butterworth filters will minimize this distortion at the expense of doing a less effective job at rejecting higher frequencies.

In general, you can always make your filter more selective by adding more poles, but this increases the complexity of your circuit (or software) and there is almost always a price for that.

In the example we just did, you can use Matlab to design the filters as follows.

```
[ num,den ] = butter(2 , 2*pi*1000 , 's');
H_butter = tf(num,den);

[ num,den ] = cheby1(2,   3,     2*pi*1000, 's');
H_cheb = tf(num,den);

[ num,den ] = ellip (2,   3, 40, 2*pi*1000, 's');
H_ellip = tf(num,den);
```

In the `butter` command, the first parameter specifies the filter order, and the 2nd parameter specifies the cutoff frequency in rads/sec. The `'s'` parameter tells Matlab to design an analog filter. In the `cheby1` command, the 2nd parameter tells Matlab how much passband ripple is acceptable (3 dB in this case). In the `ellip` command, the 3rd term tells Matlab how much stop band gain is acceptable (−40 dB in this case).

Note that these same commands can be used to design bandpass and highpass filters in Matlab. Use the help command to find out how.

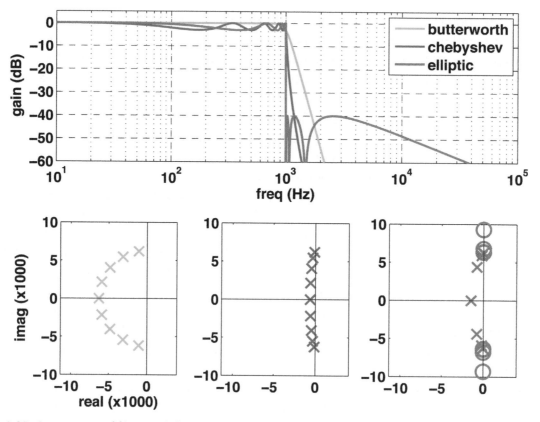

Figure 4-25: Comparison of filters with five poles and a cutoff of $f_c = 1000$ Hz.

```
help butter
help cheby1
help ellip
```

As a final note, consider the filters shown in Figure 4-25. These filters were designed exactly as in the previous example, but using five poles per filter instead of just two. As you can see, all three filters have become much more selective (i.e., better at rejecting unwanted signals) than they were in Figure 4-24. The Elliptic filter in particular is especially sharp. Notice also how the pole locations appear to be in patterns for the three filters. Butterworth poles always lie evenly spaced along an imaginary semi-circle, whereas Chebyshev poles are evenly spaced along an imaginary ellipse. Elliptic poles also fall on an ellipse but aren't evenly spaced. Note, too, the presence of the zeros for in the Elliptic filter.

Section 4.10 Summary

This chapter introduced the concept of a system, which is any entity that acts upon a signal. Electronic circuits such as amplifiers and filters are examples of systems. Systems that are linear and time invariant are most common and have the following properties.

$$c_1 x_1(t) + c_2 x_2(t) \Rightarrow c_1 y_1(t) + c_2 y_2(t)$$

$$x(t-\tau) \Rightarrow y(t-\tau)$$

There are many equivalent ways to describe a system. One method is to solve for its differential equation. In theory, the differential equation can be solved in order to determine the system output in response to any input. However, in practice, the math is often cumbersome and Fourier methods prove easier to apply. Taking the Fourier Transform of the differential equation leads to the system's transfer function, which is defined as

$$H(j\omega) = \frac{Y(j\omega)}{X(j\omega)}$$

Alternately, the system output can be expressed as

$$Y(j\omega) = H(j\omega) X(j\omega)$$

The beauty of the Fourier Transform is that it takes an unwieldy differential equation and reduces it to simple algebra in the frequency domain.

If an impulse is applied to a system, the system's response is known as the impulse response, $h(t)$. The Fourier Transform of the impulse response is always the transfer function $H(j\omega)$.

The Laplace Transform is a superset of the Fourier Transform based on complex frequencies $s = \sigma + j\omega$ instead of purely imaginary frequencies $j\omega$. The Laplace Transform allows us to solve for a system's zeros and poles, which are the roots of the transfer function's numerator and denominator, respectively. Knowing the pole and zero locations facilitates a simple method of sketching the Bode plot: zeros increase the slope of the magnitude response by 20 dB/decade, whereas poles decrease the slope by 20 dB/decade.

There are many different families of signal filters. Some, like Butterworth filters, have flat-gain profiles in the passband but relatively poor performance in the stop band. Others, like Chebyshev and Elliptic filters, exhibit passband gain ripple but make up for it with superior performance in the stop band.

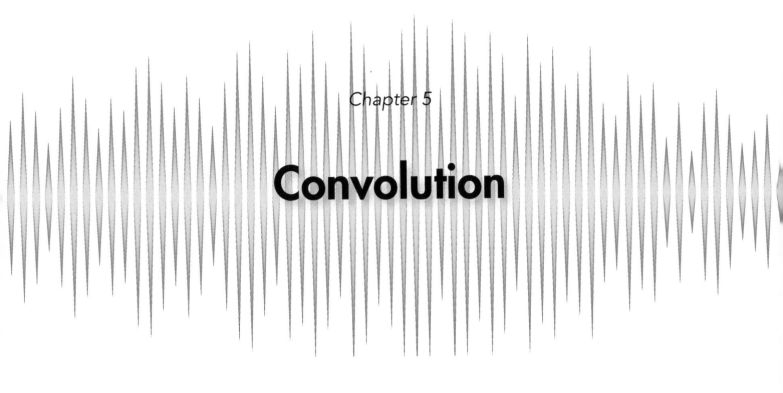

Chapter 5

Convolution

Section 5.1　Derivation

In Chapter 4, we learned how to predict the output of a system using frequency-domain analysis. Specifically, Equation [4-27] stated $Y(j\omega) = H(j\omega)X(j\omega)$, meaning that in the frequency domain, system output can be predicted using the multiplication function to combine the system characteristics with the input signal. However, Section 4.5 also suggested the possibility of a method for determining the system output in the time domain. That process is called *convolution*. We'll start by stepping through the theory behind convolution. This will lead us to the *convolution integral*, which is the formal definition of convolution. Following that, we'll develop a series of examples so we can actually practice computing the convolution integral.

5.1.1　Impulse Response

Every system has an *impulse response h(t)*, which is defined as the system's output in response to an impulse input (i.e., $x(t) = \delta(t)$). Figure 5-1 below shows an example of an impulse input and the system's impulse response. This particular impulse response happens to belong to the same first-order low-pass filter studied extensively in Chapter 4. Other filters will have impulse responses that look different, but that doesn't change anything about the underlying concepts.

　　The next step is to think about what the output would be if we modified the impulse slightly. For example, if the input impulse was delayed by τ units, then we would expect that the filter output would just be the same $h(t)$, only also delayed by τ units. Specifically, if $x(t) = \delta(t-\tau)$, then $y(t) = h(t-\tau)$.

　　Another way to modify our impulse is to make it larger or smaller by scaling it by a constant. If we modify our input by multiplying it by the constant c (i.e., $c = 2$ means doubling the size of the impulse, etc), then the output should be $h(t)$ scaled by the same constant c. Specifically, the system response to $x(t) = c \cdot \delta(t)$ is $y(t) = c \cdot h(t)$.

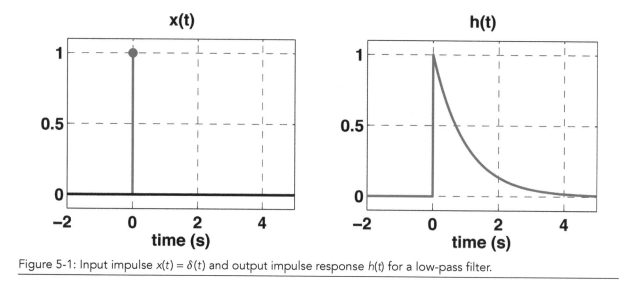

Figure 5-1: Input impulse $x(t) = \delta(t)$ and output impulse response $h(t)$ for a low-pass filter.

By combining the time-shift property and the scaling property, we arrive at the general rule that

$$x(t) = c \cdot \delta(t - \tau) \Leftrightarrow y(t) = c \cdot h(t - \tau) \qquad [5\text{-}1]$$

5.1.2 Sum of Scaled- and Shifted-Impulse Responses

Now that we know how the system will respond to any scaled or shifted impulse, let's think about how the system might respond to a more general input, such as the signal $x(t)$ shown in Figure 5-2. Although Figure 5-2 shows a signal $x(t)$ that is continuous in time, we can think of it as an infinite sum of scaled and shifted impulses. For example, in the figure we have shown a specific value of time, namely $t = \tau$. At that time, the value of the function is $x(t) = x(\tau)$. This means that, for that specific point in time, we can think of the signal as being an impulse with a delay of τ and scaled by the value $x(\tau)$, specifically

$$x(\tau) \cdot \delta(t - \tau) \qquad [5\text{-}2]$$

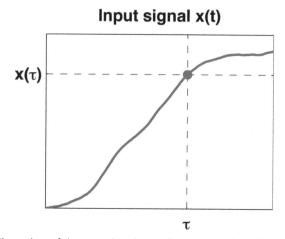

Figure 5-2: Input signal $x(t)$. The value of the signal is shown for a particular value of time, namely $t = \tau$.

And as we learned in the previous section, the filter output in response to this scaled and shifted impulse will be an impulse response that is scaled and shifted by the same amount, namely

$$x(\tau) \cdot h(t-\tau) \qquad \text{[5-3]}$$

The final step is to realize that the input signal $x(t)$ is actually equal to the sum of an infinite number of scaled and shifted impulses. In other words, if we vary τ over all values of time and sum the resulting impulses, we get an unusual (but very helpful) definition of $x(t)$.

$$x(t) = \int_{\tau} x(\tau)\delta(t-\tau)d\tau \qquad \text{[5-4]}$$

If each one of those infinite scaled and shifted impulses produces a corresponding scaled and shifted impulse response, then we can expect that the filter output y(t) will be given by

$$y(t) = \int_{\tau} x(\tau)h(t-\tau)d\tau \qquad \text{[5-5]}$$

Equation [5-5] is the *convolution integral* and is the formal definition of convolution. It tells us precisely how to calculate the filter output $y(t)$ in the time domain given the filter input $x(t)$ and the filter's impulse response $h(t)$. Note that, as shorthand, we use the $\circledast$ operator to denote convolution. Specifically

$$y(t) = x(t) \circledast h(t) \qquad \text{[5-6]}$$

Figure 5-3 demonstrates this process graphically.

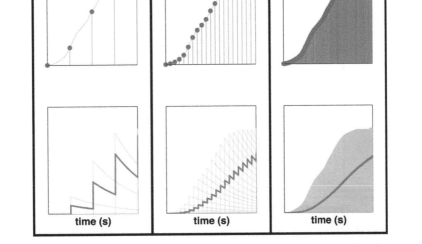

Figure 5-3: Graphical demonstration of convolution. [top] Input signal $x(t)$ (in green) and corresponding impulses (in blue). [bottom] output signal $y(t)$ is created by taking the sum (in red) of scaled and shifted impulse responses (in cyan). Left panel shows wide spacing between impulses; middle panel shows impulses more closely spaced; right panel shows zero spacing between impulses.

In Figure 5-3, the top row shows input signal $x(t)$ broken up into a series of impulses (the first column shows only five impulses; the second and third columns repeat the process with progressively more impulses). In the bottom row, the scaled-and-shifted impulses of the top row are converted into scaled-and-shifted impulse responses. These curves are then summed to produce the red signal. As the spacing between the impulses is reduced to zero (in the right panel), the sum-of-impulse-responses becomes an integral and converges to the true filter output $y(t)$. Therefore, Figure 5-3 is a graphical representation of the convolution integral.

Section 5.2   Implementing the Convolution Integral

Now that the convolution integral (Equation [5-5]) has been defined, it's worth spending some time learning how to work with it. Recall that $y(t) = \int_{\tau} x(\tau)h(t-\tau)d\tau$. Part of the trick is to realize that "time" is no longer denoted by variable t but rather by τ. For example, the integral refers to input signal $x(\tau)$ instead of $x(t)$. This doesn't actually change $x(t)$ at all, it just means that the time axis will change variable names from t to τ.

5.2.1 Computing $h(t-\tau)$

The $h(t-\tau)$ term can be confusing for Signals students and requires some explanation. Let's start with some arbitrary impulse response $h(t)$ like the one shown to the left in Figure 5-4 and work through the steps necessary to create $h(t-\tau)$.

The first step is to convert the time axis to τ. This doesn't actually change the shape of the impulse response, but it *does* change the name of the function from $h(t)$ to $h(\tau)$ (see the right side of Figure 5-4).

The next step is to replace $h(\tau)$ with $h(-\tau)$. As you should remember from pre-calculus, this procedure should simply flip $h(\tau)$ across the y-axis, as shown in the left plot of Figure 5-5.

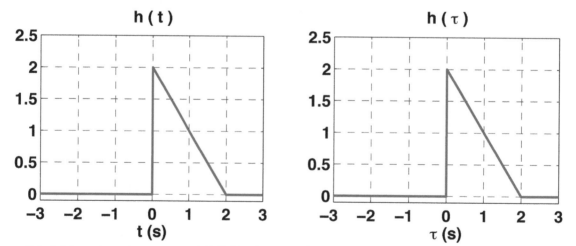

Figure 5-4: Arbitrary impulse response $h(t)$ [left]. Changing the time-axis to T [right] doesn't change the shape of the signal at all.

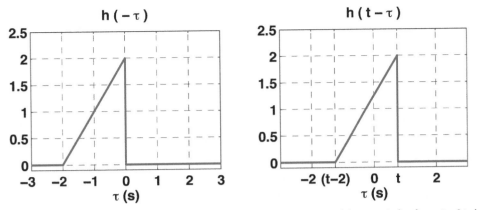

Figure 5-5: Changing $h(\tau)$ to $h(-\tau)$ [left] flips the signal across the y-axis. Adding a shift of t units [right] shifts the flipped impulse response to the right and produces $h(t-\tau)$.

The final step is to replace $h(-\tau)$ with $h(t-\tau)$. As shown in the right plot of Figure 5-5, this has the effect of shifting the flipped $h(\tau)$ to the right by t units. Therefore, to summarize, the $h(t-\tau)$ term from the convolution integral requires taking $h(t)$, flipping it, and then shifting to the *right* by t units.

From a mathematical perspective, we can generate an equation for $h(t-\tau)$ by starting with the equation for $h(t)$ and then replacing the t with $t-\tau$. For example, suppose the impulse response $h(t)$ is given by the expression $h(t) = 2-t$. The expression for $h(t-\tau)$ is therefore $h(t-\tau) = 2 - (t-\tau) = 2-t+\tau$.

5.2.2 Examples

Example 5-1

The convolution integral is computed by taking $x(\tau)$ and $h(t-\tau)$, multiplying them together, and then integrating over all τ. That procedure will give you the value of $y(t)$ for whatever value of t you used to shift the impulse response.

Let's go through a complete example to demonstrate how all this works. Suppose $x(t)$ and $h(t)$ are as given in Figure 5-6 (left and middle). These correspond to a step input $x(t)$ and the impulse response of a first-order low-pass filter $h(t)$.

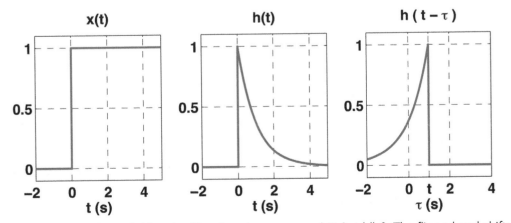

Figure 5-6: An input signal $x(t)$ [left] and a filter impulse response $h(t)$ [middle]. The flipped and shifted $h(t-\tau)$ [right] is also shown.

The equations for our functions are $x(t) = 1$ and $h(t) = e^{-t}$ for t > 0 and zero otherwise. This means that the equation for $h(t-\tau)$ is

$$h(t-\tau) = e^{-(t-\tau)} = e^{-t}e^{\tau} \qquad [5\text{-}7]$$

The right-hand plot in Figure 5-6 shows the flipped and shifted $h(t-\tau)$. The signal has been shifted to the right by t units. Note that t can be negative (negative t would just mean a shift to the left instead of to the right).

We can now begin the process of applying the convolution integral. We start by picking some value of t, say $t = -1$. The left column in Figure 5-7 shows the convolution integral graphically for $t = -1$. The top row shows us $x(\tau)$, which is just our step input. The middle row shows us $h(t-\tau)$ for $t = -1$. Notice that this results in a left-hand shift of one unit. Finally, the bottom row shows the product of $x(\tau)h(t-\tau)$. In the case of $t = -1$, we can see that the product will be zero for all values of τ.

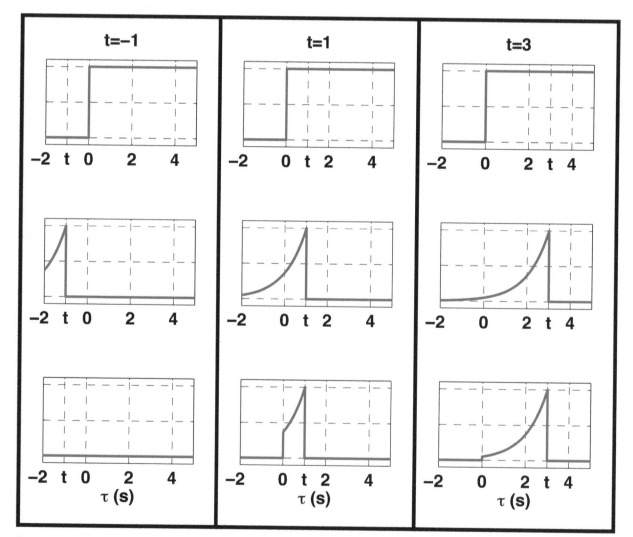

Figure 5-7: Graphical depiction of the convolution of the signals shown in Figure 5-6. [top] Input signal $x(\tau)$; [middle] flipped and shifted impulse response $h(t-\tau)$; [bottom] The product $x(\tau)h(t-\tau)$. Each column represents the convolution for a different value for t.

The final step of the convolution integral is to integrate $x(\tau)h(t-\tau)$ over all τ. By inspection, we can easily see that $y(t=-1)=0$.

That seems like a lot of work to get $y(t)$ for just one value of t, and it is! But if we are clever, we can milk the left column of Figure 5-7 for some more answers. If we look closely at the left column of Figure 5-7, we can see that our answer won't change at all as long as $t<0$. This is because, for all negative t, the flipped and shifted $h(t-\tau)$ plot won't overlap at all with $x(t)$, and therefore their product will be zero over all time. With this in mind, we can state that

$$y(t)=0 \text{ for all } t<0 \tag{5-8}$$

Next let's try to solve for $y(t)$ for some positive value of t, say $t=1$. The middle column of Figure 5-7 lays out how the convolution integral will be constructed. The top row shows $x(\tau)$, which is exactly the same as it was in the $t=-1$ case. The middle row shows $h(t-\tau)$, which is the flipped $h(t)$ shifted to the right by one unit. Finally the bottom row shows the product $x(\tau)h(t-\tau)$. We can see that the product is non-zero only on the range $0 \le t < 1$. Therefore our integral has to be

$$y(t=1)=\int_0^1 1\cdot e^{-1}e^{\tau}\,d\tau \tag{5-9}$$

This equation arises because $x(t)=1$ and $h(t-\tau)=e^{-t}e^{\tau}$ for $t>0$ (see Equation [5-7]). Solving the integral isn't so bad.

$$y(t=1)=\int_0^1 1\cdot e^{-1}e^{\tau}\,d\tau$$

$$=e^{-1}\int_0^1 e^{\tau}\,d\tau$$

$$=e^{-1}\left[e^{\tau}\right]_0^1$$

$$=e^{-1}\left[e^1-1\right]$$

$$=1-e^{-1}\approx 0.632 \tag{5-10}$$

Of course this integral gives us only the value of $y(t)$ for a single value of t, namely $t=1$. What about all the other values of t? The third column of Figure 5-7 shows another example that applies when $t=3$. Since we are clever, we notice that there is virtually nothing different between the $t=1$ and $t=3$ cases. In fact, the drawings in the 2nd and 3rd columns are basically the same, save for the actual value of t. This means that we could write one integral that would apply to all positive values of t. This would work because, for all $t>0$, the integral is the same (as evidenced by 2nd and 3rd columns of Figure 5-7). The only issue is that in solving Equation [5-9], instead of integrating from 0 to 1, we'd integrate from 0 to t.

$$y(t)=\int_0^t 1\cdot e^{-t}e^{\tau}\,d\tau$$

$$= e^{-t} \int_0^t e^{\tau} \, d\tau$$

$$= e^{-t} \left[e^{\tau} \right]_0^t$$

$$= e^{-t} \left[e^t - 1 \right]$$

$$= 1 - e^{-t} \tag{5-11}$$

Combining Equations [5-8] and [5-11], we arrive at a definition for $y(t)$ that applies over all values of time.

$$y(t) = \begin{cases} 1 - e^{-t} & t > 0 \\ 0 & t \leq 0 \end{cases} \tag{5-12}$$

The plot of $y(t)$ is shown in Figure 5-8.

You might recall that in Section 4.1, we computed the step response to a first-order low-pass filter by hand by solving the differential equation. Happily, we've just arrived at exactly the same answer here using convolution!

Example 5-2
Convolve the two signals $x(t)$ and $h(t)$ shown in Figure 5-9.

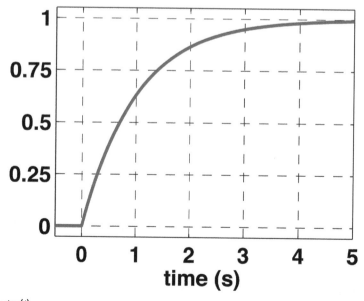

Figure 5-8: Filter output $y(t)$.

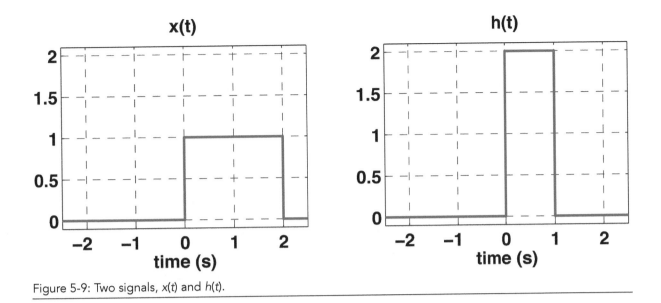

Figure 5-9: Two signals, x(t) and h(t).

As in the previous example, we start by creating plots of $x(\tau)$ and $h(t-\tau)$. The $x(\tau)$ plot is trivial because it is the same as $x(t)$ with just the axis name changed. The $h(t-\tau)$ plot requires taking $h(t)$, flipping it across the y-axis, and then shifting it by t units, keeping in mind that positive values of t produce right shifts. Figure 5-10 shows the two terms, ready for multiplying and integrating. Note that $h(t-\tau)=2$ when $t-1<\tau\leq t$ and zero otherwise.

To convolve, we have to multiply $x(\tau)$ with $h(t-\tau)$ and integrate the results over all τ. What makes this a little tricky is that, depending on the value of t, our integral can take a number of forms. Figure 5-11 shows the various cases for this problem.

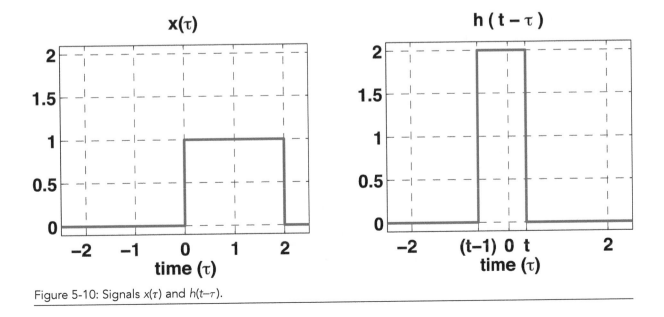

Figure 5-10: Signals $x(\tau)$ and $h(t-\tau)$.

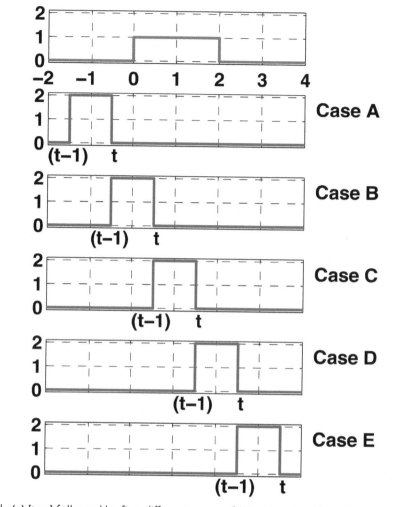

Figure 5-11: Signal $x(\tau)$ [top] followed by five different cases of $h(t-\tau)$, each with a different value of t.

The top plot in Figure 5-11 is $x(\tau)$. The five plots below it (Cases A through E) show different versions of $h(t-\tau)$, each for different values of t. For example, Case A applies whenever $t \le 0$. We know that all those values of t should be lumped into the same case because they all correspond to the situation where there is no overlap between the rectangle shapes of $x(\tau)$ and $h(t-\tau)$.

In general, we have five cases, which are detailed in Table 5-1.

Table 5-1: For Example 5-2, time t can be divided up into five distinct ranges. Each range of times produces its own integral.

Case	Range	Notes
A	$t \le 0$	No overlap between x and h
B	$0 < t \le 1$	Partial overlap between x and h
C	$1 < t \le 2$	h completely underneath x
D	$2 < t \le 3$	Partial overlap between x and h
E	$t > 3$	No overlap between x and h

Table 5-2: The complete solution to Example 5-2 broken down by Case. Each of the five Cases has its own integral that applies for all the values of t corresponding to that particular case. For example, in Case B, the solution will be $y(t) = 2t$ for all $0 < t \le 1$.

Case	Range	$y(t)$
A	$t \le 0$	$y(t) = 0$
B	$0 < t \le 1$	$y(t) = \int_0^t 2d\tau = 2t$
C	$1 < t \le 2$	$y(t) = \int_{t-1}^t 2d\tau = 2$
D	$2 < t \le 3$	$y(t) = \int_{t-1}^2 2d\tau = 6 - 2t$
E	$t > 3$	$y(t) = 0$

Each of the five cases merits its own version of the convolution integral. For example, in Cases A and E, there is no overlap between $x(\tau)$ and $h(t-\tau)$ and therefore the product $x(\tau)h(t-\tau)$ is zero over all τ. Those integrals are therefore both equal to zero. Case B is more interesting. From Figure 5-11, we can see that the non-zero product of $x(\tau)$ and $h(t-\tau)$ starts at $\tau = 0$ and ends at $\tau = t$. Therefore the integral is $\int_0^t (1) \cdot (2) d\tau = 2t$. A similar process is repeated for Cases C and D. The complete solution for this example is given in Table 5-2.

The solution to this convolution is therefore a piecewise function $y(t)$. That means that there is a different formula for $y(t)$ depending on the value of t. Figure 5-12 puts together all five Cases (as solved in Table 5-2) to create one final plot for $y(t)$. The function shown in Figure 5-12 is the solution to the convolution of the two signals shown in Figure 5-9.

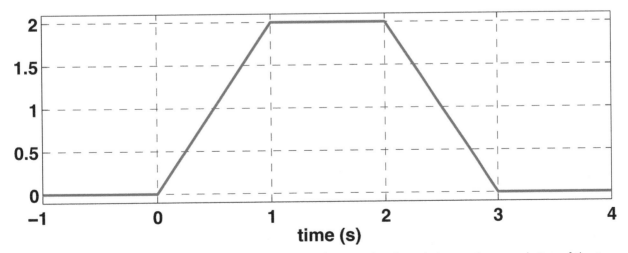

Figure 5-12: Solution for $y(t)$ as solved for in Table 5-2. This signal is the solution to the convolution of the two signals shown in Figure 5-9.

Example 5-3

In this example, we will convolve the two signals shown in Figure 5-13. Note that $x(t) = 1$ when $-1 < t \leq 1$, and $h(t) = t$ on the interval $0 < t \leq 1$.

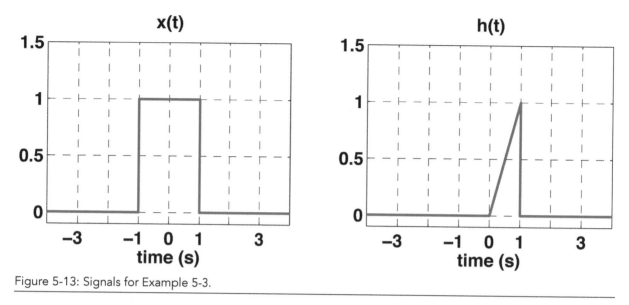

Figure 5-13: Signals for Example 5-3.

The first step is to flip and shift $h(t)$. As before, we'll do this in two steps. The first step will be to flip $h(t)$ across the y-axis (Figure 5-14, left), and the second step will be to add a right-shift of t units (Figure 5-14, right). Before we can proceed, we need a formula for $h(t - \tau)$. In general, the trick is to start with the formula for $h(t)$, and then replace every instance of t with $(t - \tau)$. In this case, we take $h(t) = t$ and wind up with $h(t - \tau) = (t - \tau)$. This function will appear prominently in our convolution integrals.

By using some trial and error, we can deduce that this example requires five cases; that is, five different integrals for five different ranges of values of t. The five cases are shown in Figure 5-15.

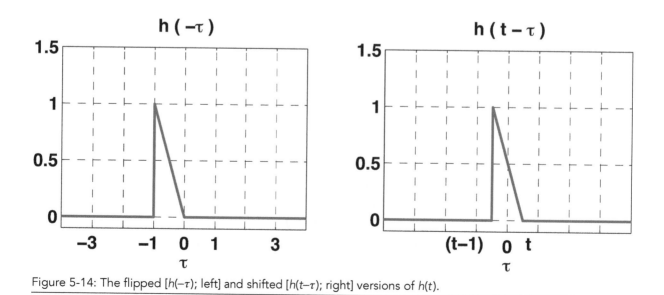

Figure 5-14: The flipped [$h(-\tau)$; left] and shifted [$h(t-\tau)$; right] versions of $h(t)$.

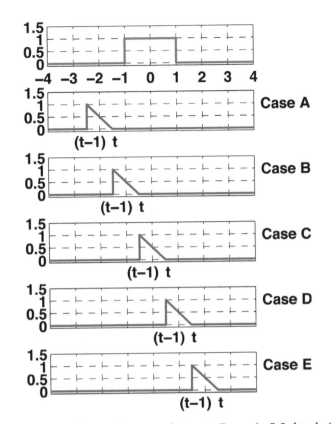

Figure 5-15: The five cases corresponding to the convolution in Example 5-3. [top] $x(\tau)$; [bottom A-E] $h(t-\tau)$ for different values of t.

As in Example 5-2, we see that a separate integral has to be applied for each of the five cases. We can solve Cases A and E by inspection by noting that there is no overlap between $x(\tau)$ and $h(t-\tau)$, which means that the integrals are both zero. The five cases and their respective integrals are summarized in Table 5-3.

Note that in each of the convolution integrals summarized in Table 5-3, $x(\tau)$ has been substituted with "1", and $h(t-\tau)$ has been substituted with "$t-\tau$".

Table 5-3: Example 5-3: Five cases, their time ranges, and their respective convolution integrals.

Case	Range	$y(t)$
A	$t \le -1$	$y(t)=0$
B	$-1 < t \le 0$	$y(t)=\int_{-1}^{t}(t-\tau)d\tau=\dfrac{1}{2}t^2+t+\dfrac{1}{2}$
C	$0 < t \le 1$	$y(t)=\int_{t-1}^{t}(t-\tau)d\tau=\dfrac{1}{2}$
D	$1 < t \le 2$	$y(t)=\int_{t-1}^{1}(t-\tau)d\tau=-\dfrac{1}{2}t^2+t$
E	$t > 2$	$y(t)=0$

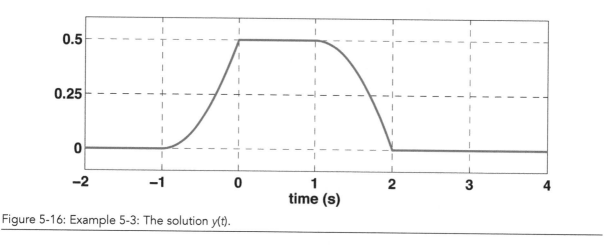

Figure 5-16: Example 5-3: The solution $y(t)$.

Figure 5-16 shows the piecewise function $y(t)$ as solved for in Table 5-3. It is interesting to note that when $0 < t \le 1$, the function is constant. This corresponds to Case C, where $h(t-\tau)$ was completely underneath $x(\tau)$. In that case, the integral does not change at all, regardless of the value of t.

Example 5-4
Figure 5-17 shows the two signals that will be convolved for Example 5-4. In this case, $h(t) = 1-t$ and therefore

$$h(t-\tau) = 1-(t-\tau)$$
$$= 1-t+\tau \qquad\qquad [5\text{-}13]$$

As in the previous examples, five cases are needed to solve the problem. These are summarized and solved in Table 5-4.

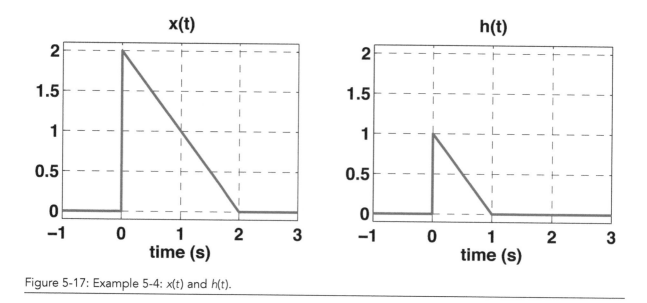

Figure 5-17: Example 5-4: $x(t)$ and $h(t)$.

Table 5-4: Example 5-4: Cases and convolution integrals.

Case	Range	$y(t)$
A	$t \leq 0$	$y(t) = 0$
B	$0 < t \leq 1$	$y(t) = \int_0^t (2-\tau)(1-t+\tau)\,d\tau = \frac{1}{6}t^3 - \frac{3}{2}t^2 + 2t$
C	$1 < t \leq 2$	$y(t) = \int_{t-1}^t (2-\tau)(1-t+\tau)\,d\tau = -\frac{t}{2} + \frac{7}{6}$
D	$2 < t \leq 3$	$y(t) = \int_{t-1}^2 (2-\tau)(1-t+\tau)\,d\tau = -\frac{1}{6}t^3 + \frac{3}{2}t^2 - \frac{9}{2}t + \frac{9}{2}$
E	$t > 3$	$y(t) = 0$

This example involves a good deal of unseemly calculus. Even though it is tedious, you are encouraged to do the integrals by hand. An engineer who isn't afraid of some untidy math is worth his or her weight in gold! The solution, as solved in Table 5-4, is plotted in Figure 5-18.

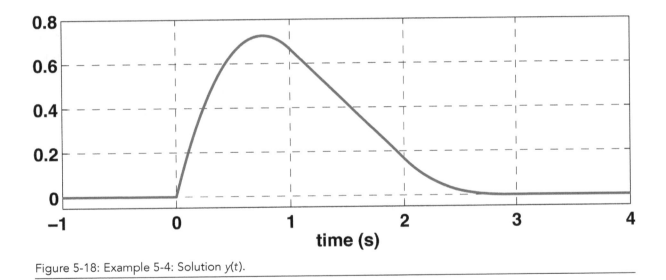

Figure 5-18: Example 5-4: Solution y(t).

Example 5-5

Figure 5-19 shows the two signals that will be convolved for Example 5-5. Note that $x(t)$ is a little complicated in this example, since we need *two* functions to describe it: $x(t) = 1+t$ when $-1 < t \leq 0$ and $x(t) = 1-t$ when $0 < t \leq 1$. Happily, $h(t)$ is just a constant and therefore $h(t-\tau) = 1$.

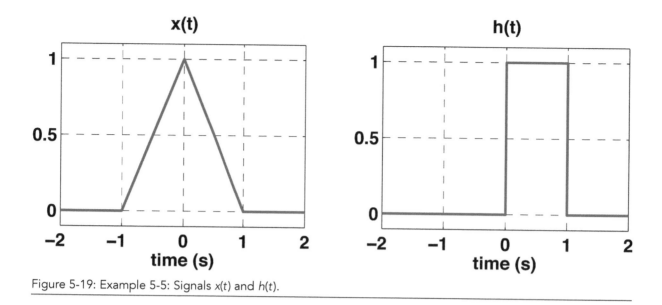

Figure 5-19: Example 5-5: Signals $x(t)$ and $h(t)$.

The steps of the convolution integral are shown in Table 5-5.

Note the issue that rises for Case C: the rectangle of $h(t - \tau)$ partially overlaps the left part of $x(\tau)$ and partially the right part of $x(\tau)$. This means that we are obliged to break down our integral (which would have been $\int_{t-1}^{t} x(\tau)h(t - \tau)d\tau$) into *two* integrals. The first integral handles the left part of $x(\tau)$ and therefore operates on the range $(t - 1) \rightarrow 0$, whereas the second integral handles the right part of $x(\tau)$ and therefore operates on the range $0 \rightarrow t$. The entire integral is described in Table 5-5.

Figure 5-20 shows the solution for Example 5-5. Note the symmetry, which is something we could have anticipated, considering that $x(t)$ and $h(t)$ are both symmetric signals themselves.

Table 5-5: Example 5-5: Cases and convolution integrals.

Case	Range	$y(t)$
A	$t \le -1$	$y(t) = 0$
B	$-1 < t \le 0$	$y(t) = \int_{-1}^{t}(1+\tau)d\tau = \frac{1}{2}t^2 + t + \frac{1}{2}$
C	$0 < t \le 1$	$y(t) = \int_{t-1}^{0}(1+\tau)d\tau + \int_{0}^{1}(1-\tau)d\tau = -t^2 + t + \frac{1}{2}$
D	$1 < t \le 2$	$y(t) = \int_{t-1}^{1}(1-\tau)d\tau = \frac{1}{2}t^2 - 2t + 2$
E	$t > 2$	$y(t) = 0$

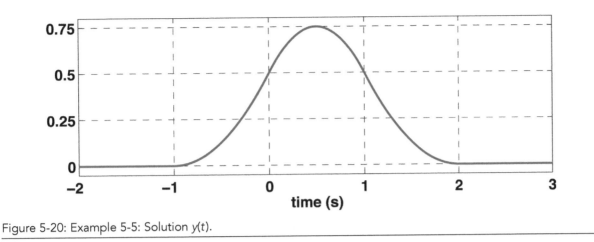

Figure 5-20: Example 5-5: Solution $y(t)$.

Section 5.3 Convolution Properties

Now that we've had some practice computing the convolution integral, let's look at some of its properties. These properties are valuable because they will help build intuition. In many cases it will be useful to use intuition to quickly infer the basic shape of a convolution problem; a computer can always be used to do the actual convolution if that level of detail is truly critical.

5.3.1 Arithmetic Properties

Convolution is commutative, associative, and distributive. The commutative property means that the order of the functions $x(t)$ and $h(t)$ doesn't matter. Stated mathematically

$$y(t) = \int_\tau x(\tau)h(t-\tau)d\tau = \int_\tau h(\tau)x(t-\tau)d\tau \qquad \text{[5-14]}$$

Or perhaps a bit more concisely

$$y(t) = x(t) \circledast h(t) = h(t) \circledast x(t) \qquad \text{[5-15]}$$

This is an incredibly useful property because many convolution problems can be made easier by switching $x(t)$ and $h(t)$. Example 5-1 could have been greatly simplified by making this switch because the $h(t-\tau)$ term would simply have been the constant 1.

The associative property means we can change the grouping of terms without affecting the result.

$$x(t) \circledast [y(t) \circledast z(t)] = [x(t) \circledast y(t)] \circledast z(t) \qquad \text{[5-16]}$$

The distributive property means that if we convolve one signal with a sum of two signals, the result is the sum of the two individual convolutions.

$$x(t) \circledast [y(t) + z(t)] = x(t) \circledast y(t) + x(t) \circledast z(t) \qquad \text{[5-17]}$$

5.3.2 Identity Function

For the convolution operator, the impulse function $\delta(t)$ is the identity function. This means that any function convolved with an impulse is unchanged.

$$x(t) \circledast \delta(t) = x(t) \tag{5-18}$$

5.3.3 Time Shift and Scaling

Convolution is linear and time invariant in the sense that if either of the two functions $x(t)$ or $h(t)$ are scaled or shifted, their convolution is scaled and shifted accordingly.

$$\text{If } x(t) \circledast h(t) = y(t)$$

$$\text{then } c_1 x(t - T_1) \circledast c_2 h(t - T_2) = c_1 c_2 y(t - T_1 - T_2) \tag{5-19}$$

A particularly useful property arises by combining the time-shift property with the identity function: convolving with a shifted impulse simply shifts your signal.

$$x(t) \circledast \delta(t - T) = x(t - T) \tag{5-20}$$

5.3.4 Start and Stop Times

In each of the examples so far in this chapter, we could have easily predicted when the output signal $y(t)$ would have started and stopped (i.e., where it is non-zero). In Example 5-2, $x(t)$ was non-zero on the range $(0, 2]$ and $h(t)$ was non-zero on $(0, 1]$. If we add the two start times and the two end times, we discover that the solution $y(t)$ must be defined on the range $(0+0, 2+1] \Rightarrow (0, 3]$, which of course is exactly what the convolution integral demonstrated. This pattern always applies, even in cases where one (or both) of the input signals extend to infinity as in Example 5-1. To summarize,

$$\text{If } x(t) \text{ is non} - \text{zero on the range } (a_x, b_x]$$

$$\text{and } h(t) \text{ is non} - \text{zero on the range } (a_h, b_h]$$

$$\text{then } y(t) \text{ is non} - \text{zero on the range } (a_x + a_h, b_x + b_h] \tag{5-21}$$

5.3.5 Multiplication and Convolution

This final property of convolution is perhaps the most interesting and certainly the most useful. If you refer back to Figure 4-10, you will notice that multiplication and convolution are related operators when it comes to signals and systems. Equation [4-27] showed that we can calculate the output signal in the frequency domain by multiplying the Fourier Transform of the input signal with the transfer function, i.e., $Y(j\omega) = X(j\omega)H(j\omega)$. What we've learned in this chapter is that the equivalent time-domain operation is convolution. Specifically, if you take the Fourier

Transform of Equation [5-6], the $x(t)$ becomes $X(j\omega)$, the $h(t)$ becomes $H(j\omega)$, the $y(t)$ becomes $Y(j\omega)$ and the convolution operator $\circledast$ becomes multiplication.

$$y(t) = x(t) \circledast h(t) \Leftrightarrow Y(j\omega) = X(j\omega)H(j\omega) \qquad [5\text{-}22]$$

As signal engineers like to say, "convolution in time is multiplication in frequency." Amazingly, the converse of this property is also true, namely that multiplication in time is equivalent to convolution in frequency!

$$y(t) = x_1(t)x_2(t) \Leftrightarrow Y(j\omega) = X_1(j\omega) \circledast X_2(j\omega) \qquad [5\text{-}23]$$

Yes, you can convolve two signals in the frequency domain using exactly the same steps outlined in this chapter! Simply replace "time" with "frequency" and proceed as normal.

Section 5.4   Applications of Convolution

In this section, we demonstrate how convolution, and specifically the properties outlined in Section 5.3, can be used to develop some incredibly powerful signal-processing applications.

5.4.1 Signal Length

Consider some sinusoid of infinite duration as shown in Figure 5-21. For illustrative purposes, we use a cosine with $f = 1k$ Hz. As we've learned, the Fourier Transform of this signal is a pair of impulses located at $f = \pm 1\text{kHz}$.

   An interesting (and very important) issue that signal engineers must consider is what the Fourier Transform of $x(t)$ looks like when the cosine is not of infinite duration. In other words, what happens if we truncate the cosine? This is important because, from a practical perspective, every cosine in the real world is of finite duration: if that weren't true you'd have to wait *forever* to create and/or analyze a signal. Fortunately, we can use convolution to address this issue.

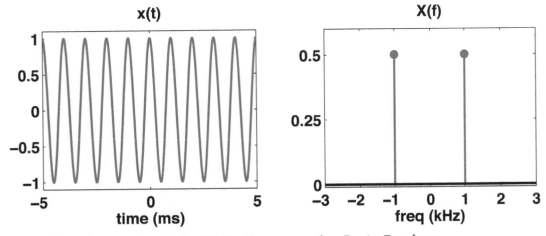

Figure 5-21: Infinite duration sinusoid ($f = 1$ kHz) with corresponding Fourier Transform.

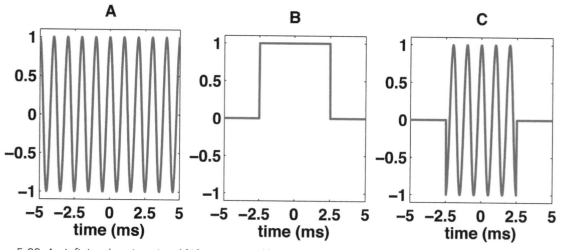

Figure 5-22: An infinite duration signal [A] is truncated by multiplying it with a finite duration square wave [B]. The resulting signal [C] is finite duration.

From a mathematical perspective, we can truncate an infinite duration signal by multiplying it with a finite duration square wave as shown in Figure 5-22.

In Figure 5-22, we arrive at the truncated signal (C) by multiplying the infinite-duration signal (A) with the finite square pulse (B). This makes sense because wherever the square pulse is zero, the product will also be zero, and wherever the square wave equals 1, the product will be just the cosine itself. In order to consider what happens in the frequency domain, we recall that multiplication in time is equivalent to convolution in frequency. Therefore, the sequence laid out in Figure 5-22 can be depicted in the frequency space as shown in Figure 5-23.

In this figure, we see that the Fourier Transform (FT) of the infinite-duration cosine [A] is convolved with the FT of the square pulse [B]. Here you should remember two important things. First, the FT of a square pulse is the sinc function, where the wider the square pulse, the narrower the sinc function, and vice versa. Secondly, when a function gets convolved with an impulse, the result is just a shifted and scaled copy of the original signal. Figure 5-23 uses this detail: the sinc function in [B] is replicated once for each of the two impulses in [A]. The FT of the finite-duration cosine is therefore shown in [C].

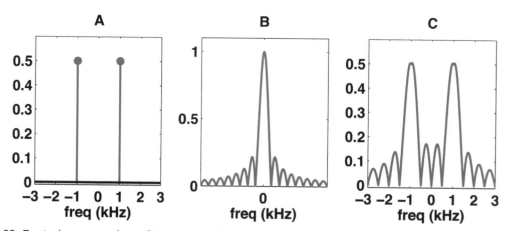

Figure 5-23: Equivalent procedure of Figure 5-22, but expressed in the Fourier domain. The FT of the input signal [A] is convolved with the FT of the square pulse [B] to arrive at the FT of the finite-duration cosine.

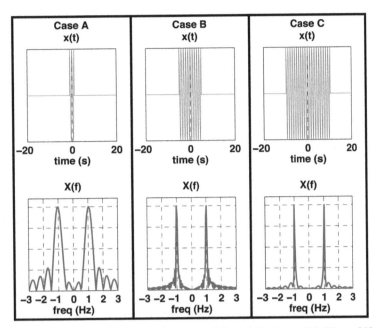

Figure 5-24: A cosine at *f = 1* kHz is truncated to a duration of 2 ms [A]; 10 ms [B]; 20 ms [C]. As the finite-duration window gets longer, the Fourier Transform (sinc pulses) gets progressively narrower.

The effect of changing the duration of the square pulse is shown in Figure 5-24 in both the time and frequency domains. Since the sinc function width is inversely proportional to the square wave width, we shouldn't be surprised to see the FT of the finite-duration signal getting sharper (i.e., more like an ideal cosine) as the finite duration gets longer (i.e., moving from [A] to [B] to [C]).

So why is this important? Let's say you were asked to study a signal and report on its frequency content. Figure 5-25 illustrates this example, in which our signal is the sum of two cosines whose frequencies are f = 100 Hz and 200 Hz.

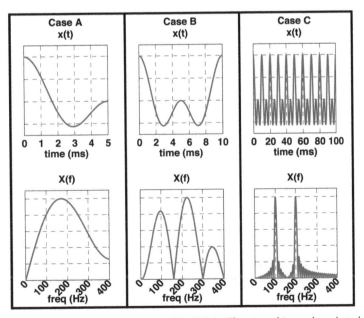

Figure 5-25: Analysis of a signal $x(t) = \cos(2\pi 100t) + \cos(2\pi 200t)$. The signal is analyzed at durations of *T = 5* ms [A]; 10 ms [B]; and 100 ms [C].

Although we know that the *ideal* Fourier Transform would contain two peaks, one each at $f = 100$ and 200 Hz, in the practical sense, that won't really be the case. Suppose that we record only our signal for $T = 5$ ms (Case A in Figure 5-25). The Fourier Transform of the very short signal shows just one large bump. If we were to analyze that bump, we would have no way of knowing that the signal we were looking at contained two distinct cosines! If instead we record the signal for a duration of $T = 10$ ms (Case B), we begin to see multiple peaks in our Fourier Transform. Only when we extend the duration of our signal to $T = 100$ ms (Case C) do we have the proper frequency resolution to definitively say that we have two peaks, one each at $f = 100$ and 200 Hz. If the time window is too small, those nice sharp peaks get broader and broader until they overlap and are eventually indistinguishable from each other (as in Case A). In general, if you want to be able to resolve frequency peaks as closely spaced as F Hz, you must use a time window whose duration is at least $T = 1/F$ seconds.

5.4.2 Amplitude Modulation

Amplitude modulation (AM) is a method of encoding signals before transmitting them; it is commonly used in radio. The equivalence of multiplication and convolution is helpful in understanding how this technology works.

Figure 5-26 shows some arbitrary signal $x(t)$ with a *bandwidth* of B Hz, which we wish to transmit over the air. A bandlimited signal is simply one with zero energy beyond its bandwidth. In principle, there is nothing to stop us from driving $x(t)$ directly into an antenna and out over the airwaves. However, a problem arises when we desire to transmit more than one signal at a time. If there are two transmitters transmitting different signals at the same time, then those signals will just add together into a useless jumble. A solution is clearly needed if we want to be able to have more than one radio station transmitting at a time. Amplitude modulation is a simple technique that allows multiple signals to occupy the same electromagnetic frequency spectrum without interfering with each other.

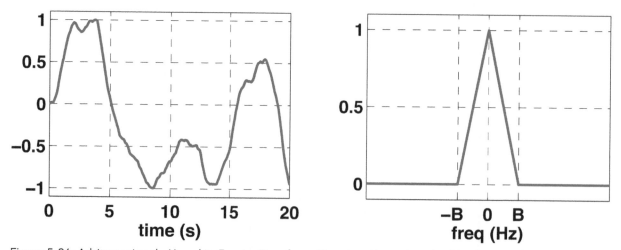

Figure 5-26: Arbitrary signal $x(t)$ and its Fourier Transform. The signal has a bandwidth of B Hz, which means all of the signal's energy is confined to frequencies less than or equal to B Hz.

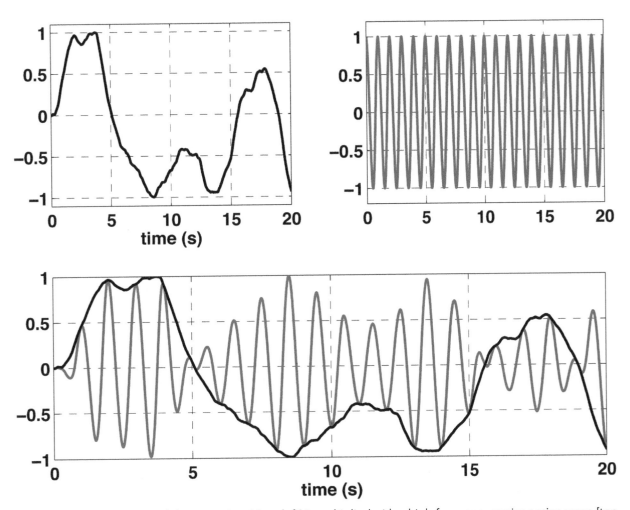

Figure 5-27: Amplitude modulation: a signal [top left] is multiplied with a high-frequency carrier cosine wave [top right] to produce an amplitude-modulated signal [bottom]. The AM signal [blue] is a cosine whose instantaneous amplitude is determined by the signal being transmitted [black].

The idea behind amplitude modulation is demonstrated in Figure 5-27. Signal $x(t)$ is multiplied with a high-frequency carrier cosine. The result is a signal that looks like a high-frequency cosine but whose instantaneous amplitude is determined by $x(t)$. As always, our goal is to think about the frequency domain in addition to just the time domain. Figure 5-28 illustrates the AM process in the frequency domain.

Since our time-domain process involved multiplying the signal $x(t)$ with a high-frequency cosine, then in the frequency domain, the equivalent process is to convolve the Fourier Transform of $x(t)$ with the Fourier Transform of the carrier, as seen in Figure 5-28. Recall that convolving with an impulse produces just scaled and shifted copies, as seen in Figure 5-28, in which the Fourier Transform of $x(t)$ has been replicated once at $-f_c$ and once again at $+f_c$.

So why is this useful? Suppose we wish to transmit multiple signals simultaneously. All we would need to do is to modulate each signal with a carrier signal at a unique frequency. For example, consider Figure 5-29. In Figure 5-29, two signals (blue and red) have each been modulated using different high-frequency carrier signals. Since each signal is band-limited to B Hz, and since the two carrier frequencies f_{c1} and f_{c2} are far enough apart, there is no overlap

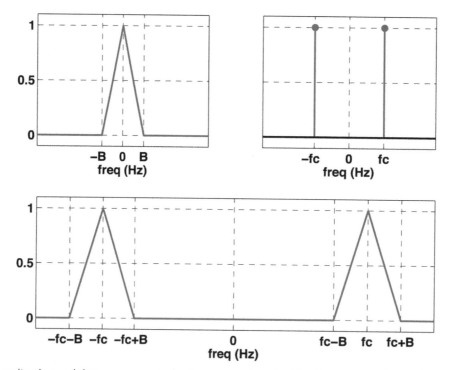

Figure 5-28: Amplitude modulation as seen in the frequency domain. The Fourier Transform of the signal [top left] is convolved with the Fourier Transform of the high-frequency carrier [top right] to produce the final AM signal [bottom].

between the signals in the frequency space, and therefore they don't interfere with each other. As long as they don't interfere, both modulated signals can co-exist in the same airspace, and both signals can be completely recovered (de-modulated).

Figure 5-29 also illustrates how to answer an important question: how closely can two carrier signals be placed? In theory, we want to pack the carrier signals as close together as possible so that we can squeeze as many channels as possible into our airwaves. Guided by the principle that we want to avoid having the red and blue signals overlap, we can see that overlap will be avoided as long as $\left(f_{c2} + B\right) \le \left(f_{c1} - B\right)$. By rearranging terms, we arrive at $f_{c1} - f_{c2} \ge 2B$. Therefore as long as the carrier frequencies are spaced by at least $2B$ Hz, there should be no overlap between channels.

The demodulation process is summarized in Figure 5-30.

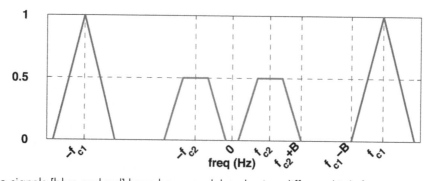

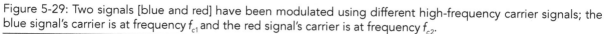

Figure 5-29: Two signals [blue and red] have been modulated using different high-frequency carrier signals; the blue signal's carrier is at frequency f_{c1} and the red signal's carrier is at frequency f_{c2}.

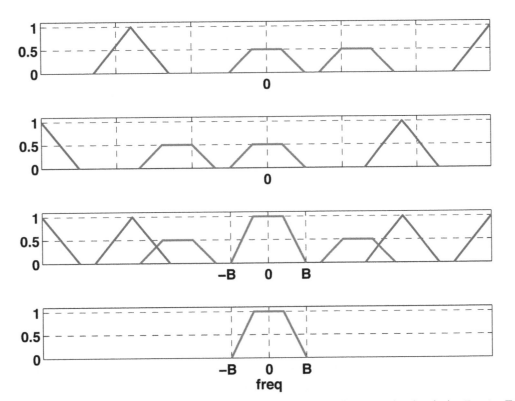

Figure 5-30: Demodulation. In the frequency domain, the received signal is convolved with the Fourier Transform of a cosine of frequency f_{c2}. This results in one copy shifted to the right by f_{c2} [top], and another copy shifted the left by f_{c2} [second row]. Their sum [third row] is then low-pass filtered with a cutoff frequency of B Hz to produce the final demodulated signal [bottom row].

Demodulation occurs in two steps. The first step is to take the received signal (which will contain all the energy in the electromagnetic spectrum, subject to the limitations of the antenna and receiver) and to multiply it with a cosine whose frequency corresponds to the frequency of the carrier signal that is to be demodulated. For example, Figure 5-30 demonstrates how to demodulate the red signal, which in Figure 5-29 was modulated at frequency f_{c2}. Since multiplication in time is convolution in frequency, the net result of the first step is to convolve the received signal with a pair of impulses at $\pm f_{c2}$ (see rows 1-3 of Figure 5-30). The second and final step is to low-pass filter the resulting signal using a cutoff frequency of B Hz. This will eliminate all signal energy above B Hz and the resulting signal will be the signal we were trying to demodulate.

In practice, AM is quite rudimentary and very susceptible to amplitude noise (which is why AM radio almost always sounds scratchy). FM (frequency modulated) radio modulates the *frequency* of a carrier signal (not its amplitude) and is therefore much more robust with respect to amplitude noise than is AM.

Section 5.5   Summary

Convolution is an operation that can be applied to two signals. In signal processing, convolution can be used to predict the system output $y(t)$ given the system input $x(t)$ and the system's impulse response $h(t)$. Convolution is defined by a convolution integral.

$$y(t) = x(t) \circledast h(t) = \int_{\tau} x(\tau) h(t-\tau) d\tau$$

Table 5-6: Properties of the convolution operator.

Commutative Property	$x(t) \circledast h(t) = h(t) \circledast x(t)$
Associative Property	$x(t) \circledast \left[y(t) \circledast z(t) \right] = \left[x(t) \circledast y(t) \right] \circledast z(t)$
Distributive Property	$x(t) \circledast \left[y(t) + z(t) \right] = x(t) \circledast y(t) + x(t) \circledast z(t)$
Identity Function	$x(t) \circledast \delta(t) = x(t)$
Time Shift and Scaling Property	$c_1 x(t - T_1) \circledast c_2 h(t - T_2) = c_1 c_2 y(t - T_1 - T_2)$
Start and Stop Times	$y(t)$ defined on the range $\left(a_x + a_h, b_x + b_h \right]$
Multiplication and Convolution	$x(t) \circledast h(t) \Leftrightarrow X(j\omega) H(j\omega)$ $x(t) h(t) \Leftrightarrow X(j\omega) \circledast H(j\omega)$

The integral essentially states that if a signal $x(t)$ can be thought of as a series of scaled and shifted impulses, then the system output must be a corresponding series of scaled and shifted impulse responses.

Convolution is characterized by the properties shown in Table 5-6.

Chapter 6

Discrete-Time Signals and Systems

Section 6.1 Discrete-Time Signals

So far we have considered only *continuous-time* (CT) signals. CT signals represent real-world signals, because in real life, every signal is continuous (e.g., audio). If you've ever listened to audio on a cassette tape or an LP record, those are CT recordings because sound is defined continuously in those media formats. An example of an arbitrary CT signal $x(t)$ is shown in Figure 6-1.

While the real-world operates in CT, computers do not. Computers are capable only of saving and manipulating lists of numbers. So if you want to capture a CT signal such as a sound and put it into your computer, you must break that sound into a finite list of numbers. When a signal is broken down into a finite list of values, it is said to be *discrete time* (DT). Whenever you save audio files on your computer (or MP3 player or CD/DVD), you are dealing with a DT signal. DT signals are intended to capture all the information of a particular CT signal.

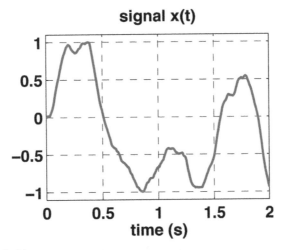

Figure 6-1: Arbitrary CT signal $x(t)$.

signal x(t)

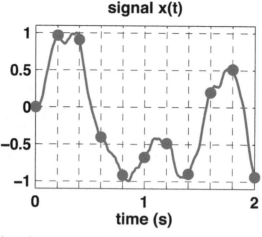

Figure 6-2: Signal x(t) with samples taken every 0.2 seconds.

6.1.1 Sampling

Figure 6-2 shows how a CT signal can be *sampled* into a DT signal. In this case, signal $x(t)$ has been sampled five times every second. We say that the *sampling rate* is $F_s = 5$ Hz or $F_s = 5$ samples/sec. In general, the reciprocal of the sampling rate F_s is the *sampling period, $T_s = 1/F_s$*, which tells us how much time elapses between samples. In this case, $F_s = 0.2 s$. Note from Figure 6-2 that there are 11 samples. If we were to number these samples (starting with zero), we would say that we have samples $n = 0, 1, \ldots, 10$. If we were then to plot just the samples by themselves without the underlying signal $x(t)$, we would get the DT signal shown in Figure 6-3.

Notice that in Figure 6-3, the x-axis no longer reads "time" but rather "samples." This is a fundamental difference between CT and DT signals. When we speak about $x[1]$, we are referring to the first sample of a DT signal. By comparison, when we speak about $x(1)$, we are talking about the value of x at time $t = 1$ second. Notice also that parentheses are used when describing CT signals whereas square brackets are used for DT signals.

signal x[n]

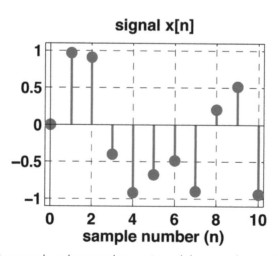

Figure 6-3: Signal x[n] shows the sample values on the y-axis and the sample number on the x-axis.

Fourier Theory tells us that any CT signal $x(t)$ can be broken down into a sum of cosines. It is therefore reasonable to start our discussion of sampling by considering only cosines. Whatever we learn through that analysis can easily be extended to any arbitrary signal by summing cosines of many frequencies, as Fourier theory has taught us.

6.1.2 Discrete-Time Frequency

Suppose we have a CT signal $x(t) = cos(2\pi ft)$ that we wish to sample at some sampling rate F_s samples per second. This means that we are taking one sample every $1/F_s$ seconds. Mathematically speaking, we are evaluating $x(t)$ only at certain times, specifically at $t = n/F_s$ where n is the "sample number" (0, 1, 2, ...). By substitution we arrive at

$$x[n] = cos(2\pi ft)\big|_{t=n/F_s}$$

$$= cos\left(\frac{2\pi fn}{F_s}\right)$$

$$= cos(\Omega n) \qquad [6\text{-}1]$$

where

$$\Omega = \frac{2\pi f}{F_s} \qquad [6\text{-}2]$$

We refer to Ω as the *discrete-time frequency* of signal $x[n]$. By simple dimensional analysis, we can see that the units of Ω must be radians per sample. Let's now illustrate the sampling process with three straightforward examples.

Example 6-1
Sample a 10 Hz cosine at $F_s = 50\,Hz$.

Solution 6-1
$$x[n] = cos(2\pi 10t)\big|_{t=n/50} = cos\left(\frac{2\pi 10n}{50}\right) = cos(0.4\pi n)$$

This implies that $\Omega = 0.4\pi$ rads/sample. Let's stop for a moment to consider what it means to have a DT frequency of 0.4π rads/sample. We know that any periodic signal repeats itself every 2π radians. In this particular case, if we had five samples, then we'd have $5 \times 0.4\pi = 2\pi$ radians. Therefore, our signal should repeat itself with a period of five samples, as shown in Figure 6-4. Therefore, "radians per sample" is a measure of how many radians each sample contributes. Generally speaking the period of a DT cosine is

$$T = \frac{2\pi}{\Omega}\ samples \qquad [6\text{-}3]$$

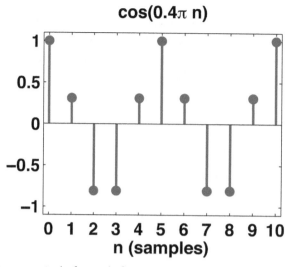

Figure 6-4: $x[n] = \cos(0.4\pi n)$ has a period of exactly five samples.

Example 6-2

Sample a 200 Hz cosine at $F_s = 1000$ Hz.

Solution 6-2

$$x[n] = \cos(2\pi 200 t)\big|_{t=n/1000} = \cos\left(\frac{2\pi 200 n}{1000}\right) = \cos(0.4\pi n)$$

Amazingly, we wind up with the exact same $x[n]$ as we did in Example 6-1. In both cases we get a cosine with frequency $\Omega = 0.4\pi$ rads/sample. Let's see how a third example goes.

Example 6-3

Sample a 5000 Hz cosine at $F_s = 25000$ Hz.

Solution 6-3

$$x[n] = \cos(2\pi 5000 t)\big|_{t=n/25000} = \cos\left(\frac{2\pi 5000 n}{25000}\right) = \cos(0.4\pi n)$$

Incredibly, we get the same DT signal in all three cases, despite the fact that we started with three very different continuous-time cosines. This is an important observation! It tells us that we can't properly interpret a discrete-time signal such as $\cos(0.4\pi n)$ unless we also know the sampling frequency F_s that was used to acquire that signal. Without knowledge of the sampling frequency, it's impossible to know whether $\cos(0.4\pi n)$ came from a 10 Hz, 200 Hz, or 5000 Hz signal. However, if we know the sampling rate, we can work backwards from the DT frequency to determine what the frequency was of the original CT signal. Rearranging Equation [6-2] we see

$$f = \frac{\Omega F_s}{2\pi} \qquad [6\text{-}4]$$

Therefore if we know $\Omega = 0.4\pi$ rads/sample and $F_s = 25000$ Hz, then we can easily see that our CT signal had to have been a cosine at frequency $f = 0.4\pi \cdot 25000 / 2\pi = 5000$ Hz.

In the next section, we'll try to determine whether or not the choice of sampling rate is important. In Figure 6-2, we used a sampling frequency of $F_s = 5$ Hz to sample $x(t)$, meaning five samples every second. What if we'd used four samples every second, or maybe only one sample every second? Would that be enough samples to accurately capture the characteristics of the signal in Figure 6-2? It should be apparent that there should be some minimum sampling rate to properly capture signal $x(t)$. The next section will explore that phenomenon and lead us to a firm rule for setting the sampling rate.

Section 6.2 Aliasing

Looking back at Figure 6-2, our intuition tells us that there must exist some sort of minimum sampling rate below which we won't have enough samples to properly capture the shape of the signal. Thinking about it further, it seems reasonable that the minimum required sampling rate should be somehow tied to the frequency content of the signal. For example, a 1 Hz cosine would probably need very few samples per second as compared to a 1000 Hz cosine.

The purpose of sampling is to capture enough points that the original CT signal could be precisely reconstructed from the DT signal. Let's look at two specific examples: one that is properly sampled and one that isn't.

Figure 6-5 shows a 200 Hz cosine (black trace). You can tell it is 200 Hz because its period is $T = 1/200 = 5$ ms. Next we sample the 200 Hz cosine at $F_s = 1000$ Hz. This means that we take 1000 samples per second, or equivalently, one sample every 1 ms. The red dots in Figure 6-5 represent these samples. Finally, we attempt to use the information in the discretized samples to reconstruct our CT signal. By fitting a cosine to the red dots, we see that the resulting CT signal (green trace) is exactly the same as our original black trace. Since the original and reconstructed signal are identical, we can say for sure that we had used a high-enough sampling rate in this example. As we learned in the previous section, the DT frequency of the samples (red dots) should be $\Omega = 2\pi 200 / 1000 = 0.4\pi$ rads/sample.

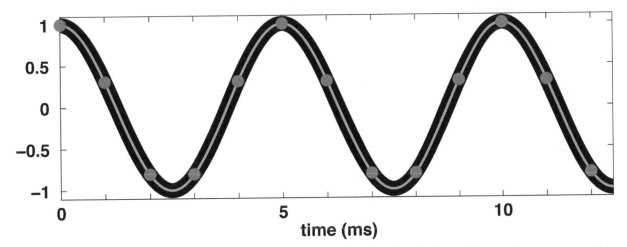

Figure 6-5: A cosine at 200 Hz [black trace] is sampled at 1000 Hz [red dots]. The red dots can be reconnected to get the reconstructed CT signal [green].

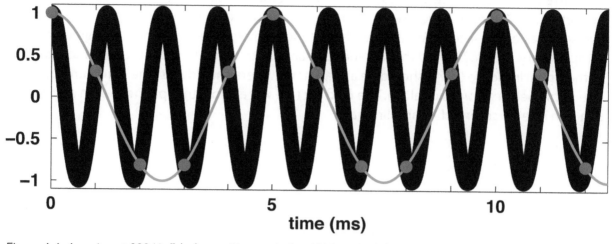

Figure 6-6: A cosine at 800 Hz [black trace] is sampled at 1000 Hz [red dots]. The red dots can be reconnected to get the reconstructed CT signal [green].

Next we'll try a case where the sampling rate is too low. We'll try to sample an 800 Hz signal at $F_s = 1000$ Hz. Figure 6-6 shows the 800 Hz signal (period $T = 1/800 = 1.25$ ms) in black. We can tell the black signal is at 800 Hz because it cycles four times in 5 ms. As before, we generate samples at $F_s = 1000$ Hz, which corresponds to one sample every 1 ms. Figure 6-6 shows the red samples that result when the 800 Hz signal is sampled every 1 ms. However, if we attempt to rebuild the CT signal by fitting a cosine to the samples, we find that the resulting CT signal is in fact very different from the one we started with. Since we were unable to reconstruct our original CT signal from the samples, we can say that our F_s was too small. Our DT frequency in this case is $\Omega = 2\pi 800/1000 = 1.6\pi$ rads/sample.

6.2.1 Calculating the Aliased Frequency

A close comparison of Figure 6-5 and Figure 6-6 shows that the reconstructed CT signals (the green ones) are *exactly* the same in both cases. In other words, it appears that $\cos(0.4\pi n) = \cos(1.6\pi n)$. Can this really be true? Consider the first five samples of each signal, which you can easily verify with a calculator, as shown in Table 6-1. It appears that the samples are exactly the same! So to recap, we sampled both a 200 Hz cosine and an 800 Hz cosine at $F_s = 1000$ Hz and wound up with exactly the same discrete-time signal, even though their DT frequencies (0.4π

Table 6-1: Samples from $\cos(0.4\pi n)$ and $\cos(1.6\pi n)$.

n	$\cos(0.4\pi n)$	$\cos(1.6\pi n)$
0	1	1
1	0.31	0.31
2	−0.81	−0.81
3	−0.81	−0.81
4	0.31	0.31

and 1.6π, respectively) are quite different. This is bad because if we look at the samples, we have no way of knowing whether they correspond to $\Omega = 0.4\pi$ (and therefore a 200 Hz cosine) or $\Omega = 1.6\pi$ (and therefore an 800 Hz cosine).

This phenomenon is called *aliasing* and it occurs because the 800 Hz signal wasn't discretized using enough samples (as evidenced in Figure 6-6). Aliasing occurs any time a CT signal is sampled with a sampling frequency that is too small: the resulting DT signal appears to have been sourced from a CT signal of much lower frequency than it actually was. In our example, the 800 Hz signal has been aliased and we have no way of knowing whether the DT signal was sourced from an 800 Hz or a 200 Hz cosine. Aliasing is bad and is generally to be avoided. For example, suppose you are trying to record someone playing notes on a piano and they strike a key that creates a tone at 800 Hz. If you sample that signal using a 1000 Hz sampling rate, then when you play back that signal the note would sound like it was 200 Hz. Your listeners would be hearing the wrong note! And that would make you a pretty lousy audio engineer. Signals that are under-sampled become aliased, which scrambles your signal and changes the information it carries.

Having defined aliasing, let's examine the math and see if we could have predicted that $\Omega = 1.6\pi$ and $\Omega = 0.4\pi$ are functionally identical. The easiest way to think about this problem is to consider the plot shown in Figure 6-7, which shows $\cos(\Omega)$ vs. Ω.

Consider the blue dot located at $\Omega = 1.6\pi$ radians. We can see by symmetry that $\cos(1.6\pi) = \cos(0.4\pi)$, and by extension $\cos(1.6\pi n) = \cos(0.4\pi n)$. Figure 6-7 makes it clear why the 800 Hz cosine ($\Omega = 1.6\pi$ rads/sample) was aliased down to 200 Hz ($\Omega = 0.4\pi$ rads/sample). Similarly, Figure 6-7 shows that a DT signal with $\Omega = 3.3\pi$ is equivalent to a DT signal of frequency $\Omega = 0.7\pi$.

6.2.2 Nyquist Theorem

A closer examination of Figure 6-7 shows that any DT frequency greater than $\Omega = \pi$ rads/sample has an equivalent value between 0 and π rads/sample. In other words, *the highest possible DT*

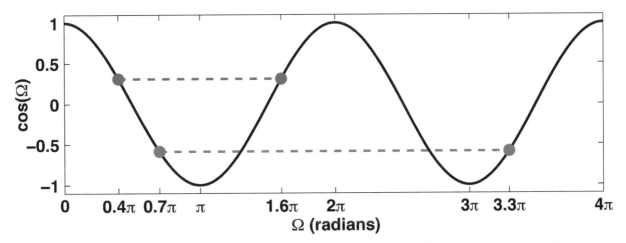

Figure 6-7: The signal $\cos(\Omega)$. The blue dots mark $\Omega = 0.4\pi$ and 1.6π radians. The red dots mark $\Omega = 0.7\pi$ and 3.3π radians.

frequency is $\Omega = \pi$ *rads/sec.* Any signal with a higher frequency has an equivalent value on the range $0 < \Omega_{eq} \leq \pi$ rads/sample.

Finally, we examine what it means for DT signals to have a maximum frequency of $\Omega = \pi$ rads/sample. Using the definition of Ω, we see that

$$\Omega_{eq} \leq \pi$$

$$\frac{2\pi f}{F_s} \leq \pi$$

$$F_s \geq 2f \qquad\qquad [6\text{-}5]$$

Finally we have arrived at our goal: a rule dictating the minimum sampling frequency for a CT signal. The rule $F_s \geq 2f$ is known as the *Nyquist Theorem* and it tells us that we must sample our CT signal using a sampling frequency of *at least* double the signal frequency. If we want to sample a 200 Hz signal, we must use a sampling rate of at least 400 Hz. Conversely, if we are using a sampling rate of 1000 Hz, we are restricted to sampling CT signals with frequencies of 500 Hz or less. Failure to abide by the Nyquist theorem will result in signals that are aliased, meaning they will appear as lower frequency signals, as shown in Figure 6-6.

6.2.3 Aliasing—Frequency Perspective

So far, we've only considered the sampling problem from the time-domain perspective. Here we'll examine the frequency perspective, and in doing so we'll validate the Nyquist Theorem. Our end result will tell us exactly what we already know (namely $F_s \geq 2f$) but the alternate derivation will lead to some good intuition on the problem in general.

Consider a CT signal $x(t)$ that is band limited between $-B < f < B$ Hz (Figure 6-8).

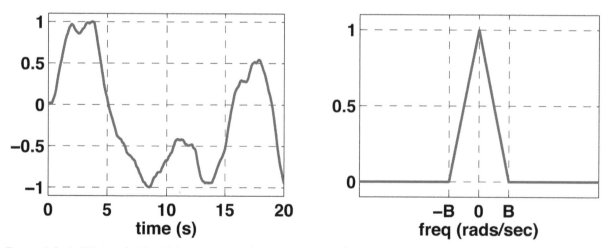

Figure 6-8: A CT signal $x(t)$ with its corresponding Fourier Transform. Note that the signal is band limited on the range $-B < f \leq B$ Hz.

Our goal is to sample $x(t)$ with a sampling rate of F_s samples per second. One way to interpret this is that we are going to take $x(t)$ and multiply it with a train of impulses called $s(t)$ that are spaced at $1/F_s$ seconds apart. This will leave us with a train of impulses whose heights [areas] are scaled according to the values of $x(t)$ at the time of the particular impulses (see Figure 6-9).

The function $x(t) \times s(t)$ is very nearly the DT signal we are aiming for. The only remaining step is to change the x-axis from time (seconds) into samples. Although this is an important mathematical step, we'll omit it here since it's a bit nuanced and doesn't add much to our intuitive understanding of aliasing from the frequency perspective.

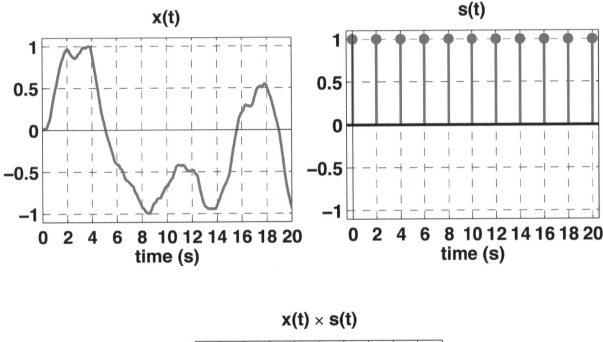

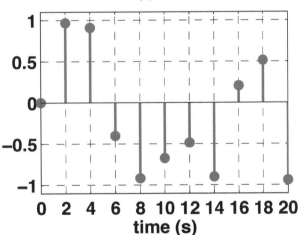

Figure 6-9: In order to sample CT signal $x(t)$, we multiply it with an impulse train where the impulses occur every $1/Fs$ seconds. In this case, the sampling rate is $F_s = 0.5$ Hz, or 2 seconds per sample. The product of these two functions is a train of impulses whose heights [areas] are scaled by the corresponding values of $x(t)$.

Recalling our Fourier Transform properties, we've learned that when we multiply two signals in the time domain, as we're doing here with $x(t)$ and $s(t)$, the equivalent operation in the frequency domain is to convolve their Fourier Transforms. Since the Fourier Transform of $x(t)$ is given (see Figure 6-8), all we need to proceed is the transform of $s(t)$. This is easily solved. Since $s(t)$ is a periodic signal (see Figure 6-9; period is $T = 1/F_s$ seconds), we can use the Fourier Series formula to solve for its frequency content.

$$A_n = \frac{1}{T} \int_T s(t) e^{-j\frac{2\pi n}{T}t} \, dt$$

$$= F_s \int_{-1/2F_s}^{1/2F_s} s(t) \cdot e^{-j2\pi n t F_s} \, dt$$

$$= F_s \int_{-1/2F_s}^{1/2F_s} \delta(t) \cdot e^{-j2\pi n t F_s} \, dt$$

$$= F_s \cdot e^{-j2\pi n \cdot 0 \cdot F_s}$$

$$= F_s \qquad\qquad\qquad\qquad [6\text{-}6]$$

Therefore, at each and every harmonic, the value of the FT is the same: F_s. The harmonic frequencies in this case are $\omega_n = 2\pi n / T = 2\pi F_s n$ rad/sec, or more conveniently, $f_n = nF_s$. It appears that the FT of the impulse train is yet another impulse train in the frequency domain where the impulses are spaced every $f = F_s$ Hz and each pulse has the exact same height: F_s. (see Figure 6-10).

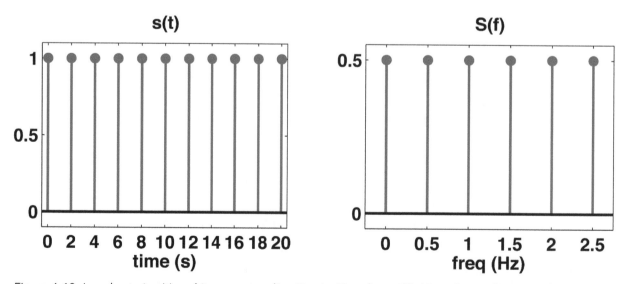

Figure 6-10: Impulse train $s(t)$ and its corresponding Fourier Transform $S(f)$. Note that in this example, the sampling rate is $F_s = 0.5$ Hz, which means the sampling period is $T = 1/0.5 = 2$ seconds. Signal $S(f)$ is an impulse train in the frequency domain, with impulses spaced every F_s Hz; each impulse has an area of value F_s.

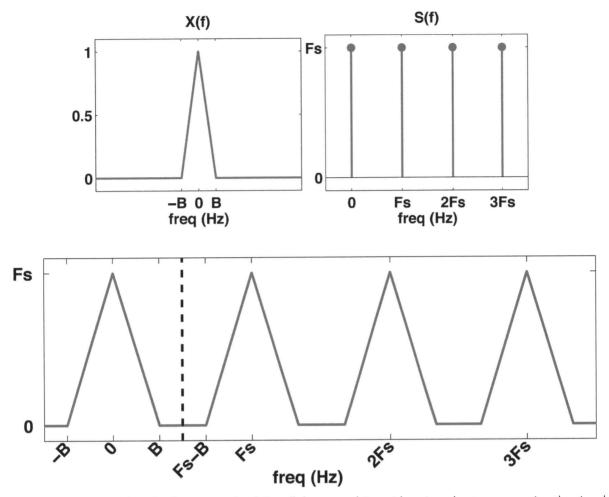

Figure 6-11: Signals $X(f)$ and $S(f)$ are convolved. Recall that convolving with an impulse means copying the signal at the impulse location and scaling by the impulse area. As long as the triangle shapes don't overlap, the information in the signal X(f) will have been perfectly captured and aliasing will be avoided. This occurs as long as $B \leq (F_s - B) \Rightarrow F_s \geq 2B$. The black dashed line shows the Nyquist frequency $F_s/2$.

Now that we have $S(f)$, all that remains is to convolve the FT of $x(t)$ with the FT of $s(t)$. Note that $S(f)$ is an impulse train, and we have learned that convolving with an impulse means copying the signal $X(f)$ at every instance of the impulse. In our case, we get the result shown in Figure 6-11.

Figure 6-11 shows the signal $X(f)$ replicated once for every impulse of $S(f)$, whose impulses are spaced every F_s. Figure 6-11 also shows us where the Nyquist Theorem comes from. It should seem reasonable that as long as the triangles aren't overlapping, the information in signal $x(t)$ will have been preserved by the sampling process. Suppose now that we decide to reduce the sampling frequency F_s Hz to some smaller value. As F_s is reduced, the triangle shapes will get closer and closer together. At the point where they begin to overlap, the original triangle shape will be altered, which we can interpret to mean that the sampling frequency is now too small to properly capture all the information in $x(t)$. In other words, *aliasing*. Figure 6-11 clearly shows that aliasing can be avoided as long as

$$B \leq F_s - B$$

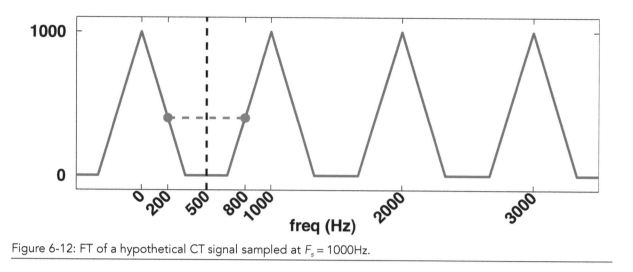

Figure 6-12: FT of a hypothetical CT signal sampled at $F_s = 1000$Hz.

$$F_s \geq 2B \qquad\qquad \text{[6-7]}$$

or equivalently

$$B \leq \frac{F_s}{2} \qquad\qquad \text{[6-8]}$$

This is the Nyquist Theorem—the sampling frequency must be at least twice the value of the highest-frequency energy in the signal in order to avoid aliasing. Or equivalently, the maximum CT frequency must be half the sampling frequency.

Figure 6-11 also demonstrates why aliased signals appear indistinguishable from lower frequencies. For example, if we modify Figure 6-11 by specifying a sampling frequency of $F_s = 1000$ Hz, we see that discretized $f = 200$ Hz signals will be indistinguishable from discretized 800 Hz ones. This is exactly the realization we observed in Figure 6-7. Figure 6-12 also clearly shows that the Nyquist frequency is 500 Hz, and that any CT signal above that frequency will alias when sampled at $F_s = 1000$ Hz.

In reality, there is no such thing as a CT signal that is purely band limited at $\pm B$ Hz. Consequently, if the signal is not band limited, Figure 6-12 makes it clear that there will always be aliasing, regardless of how large F_s is. The standard solution is to low-pass filter $x(t)$ before sampling it. Used in this context, the low-pass filter is known as an "anti-aliasing" filter. Recall that no filter is ideal, so even when $x(t)$ is anti-aliased, there will still be some resulting energy at frequencies above B Hz. Therefore, as a practical rule of thumb, the sampling frequency is usually at least 3–5 times B to minimize the worst of the aliased signals.

Section 6.3 Discrete-Time Fourier Transform

Now that we understand discrete-time frequencies and sampling, we can discuss the Discrete-Time Fourier Transform (DTFT), which is how we convert DT signals from time to frequency representations. Remember, unlike continuous-time frequencies that can range all the way to infinity ($0 \leq \omega < \infty$) rads/sec, discrete-time frequencies are unique only up to $\Omega = \pi$ rads/sample which corresponds to signals at the Nyquist rate (e.g., half the sampling frequency, or $F_s / 2$).

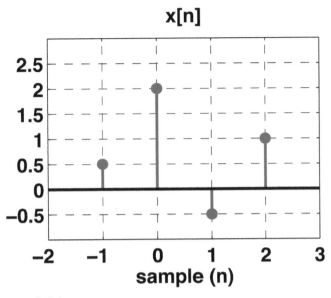

Figure 6-13: Discrete-time signal x[n].

Frequencies greater than $\Omega = \pi$ rads/sample are simply duplications of frequencies on the range $0 \le \Omega < \pi$ (see Equation [6-5]).

The Discrete-Time Fourier Transform is defined as

$$H(\Omega) = \sum_n x[n]e^{-j\Omega n} \qquad [6-9]$$

This is essentially the same definition as the Continuous-Time Fourier Transform, except the integral over time has been replaced with a sum over samples. In many respects, it is easier to calculate the DTFT because it does not require any integrals. Let's examine with an example. Suppose $x[n]$ is as shown in Figure 6-13. An alternative representation of the same signal would be to simply list the samples in a table.

n	−1	0	1	2
$x[n]$	0.5	2	−0.5	1

With the values clearly laid out in the table, we can calculate the DTFT by simply plugging numbers into Equation [6-9].

$$H(\Omega) = 0.5e^{-j\Omega(-1)} + 2e^{-j\Omega 0} - 0.5e^{-j\Omega 1} + 1e^{-j\Omega 2}$$

$$= 0.5e^{j\Omega} + 2 - 0.5e^{-j\Omega} + e^{-j2\Omega} \qquad [6-10]$$

Equation [6-10] is a *complex function* because all those complex exponentials will return complex numbers when we evaluate at various values of Ω. This means that, as with the CTFT, we must examine both the magnitude and the phase, which requires using two separate plots, as shown in Figure 6-14.

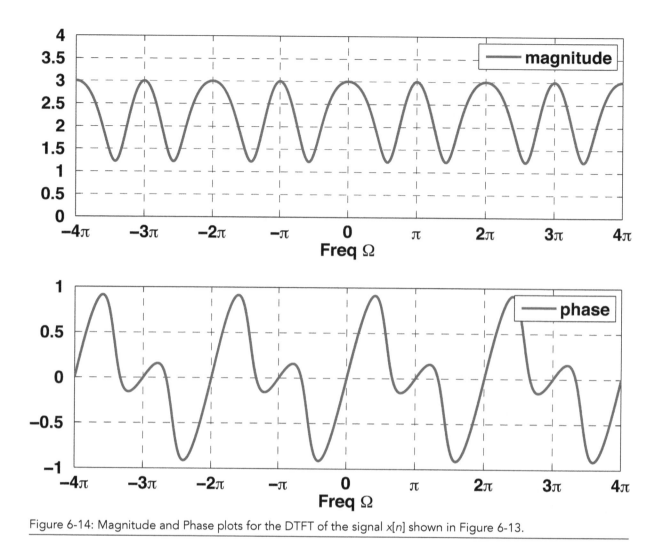

Figure 6-14: Magnitude and Phase plots for the DTFT of the signal x[n] shown in Figure 6-13.

The plot reveals several interesting things. First, the magnitude plot is even and the phase plot is odd. This should come as no surprise to us since the same properties were also true of the CTFT plots. The second interesting thing is that the DTFT is periodic with period 2π. In other words, once you get to $\Omega = \pi$, the signal starts repeating the same values from $\Omega = -\pi$. This is exactly as expected because we learned before that DT frequencies are unique only up until $\Omega = \pi$, which corresponds to the Nyquist frequency $F_s/2$. To summarize, *every* discrete-time signal has the same three properties:

1. Magnitude of Transfer Function is even
2. Phase of Transfer Function is odd
3. Transfer Function is periodic with period 2π

As a sanity check, let's try three known cosines and see if their DTFTs make sense. We'll try $x_1[n] = \cos(0\pi n)$, $x_2[n] = \cos(\pi/2n)$ and $x_3[n] = \cos(\pi n)$. In each case, we'll include eight samples in our calculation. The signal samples and their DTFT magnitude plots are shown in Figure 6-15.

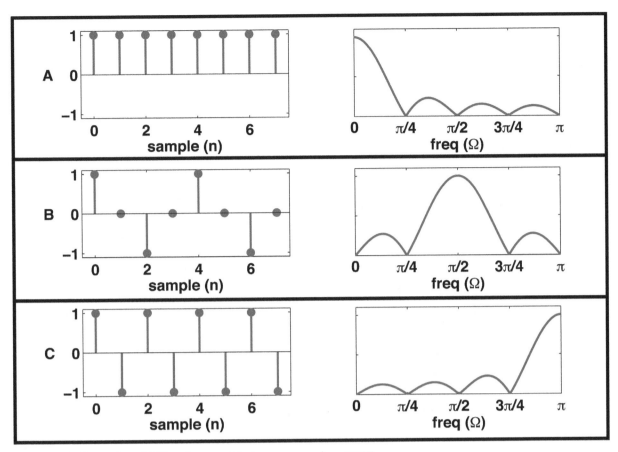

Figure 6-15: Examples of DT cosines and their corresponding DTFTs.

In case A, the DT frequency is $\Omega = 0$ rads/sample, which produces a constant signal. The corresponding DTFT, given by

$$H = 1e^{-j\Omega 0} + 1e^{-j\Omega 1} + 1e^{-j\Omega 2} + 1e^{-j\Omega 3} + 1e^{-j\Omega 4} + 1e^{-j\Omega 5} + 1e^{-j\Omega 6} + 1e^{-j\Omega 7} \qquad [6\text{-}11]$$

shows that most of the energy is located around $\Omega = 0$ rads/sample, as expected! If our DT signal had infinite duration, we would of course expect that the DTFT would look like a pure impulse at $\Omega = 0$ rads/sample. However, we are using only a finite duration signal, so for the same reasons discussed in Section 5.4.1, the DTFT is not an ideal impulse.

In case B, the frequency is $\Omega = \pi / 2$ rads/sample. As a check, we know from Equation [6-3] that the period should be $T = 2\pi / (\pi / 2) = 4$ samples, which agrees with the plot. The corresponding DTFT, given by

$$H = 1e^{-j\Omega 0} + 0e^{-j\Omega 1} - 1e^{-j\Omega 2} + 0e^{-j\Omega 3} + 1e^{-j\Omega 4} + 0e^{-j\Omega 5} - 1e^{-j\Omega 6} + 0e^{-j\Omega 7} \qquad [6\text{-}12]$$

clearly shows most of the signal energy at $\Omega = \pi / 2$ rads/sample.

Case C shows $\Omega = \pi$ rads/sample, which as we know is the highest unique DT frequency. Again, the DTFT is as expected.

Section 6.4 FIR Filters

Next we consider discrete-time systems. There are two basic types of DT filters: finite impulse response and infinite impulse response. They have different properties and applications, and both are useful for the digital engineer to understand. This section focuses on finite impulse response (FIR) filters.

Consider the following discrete-time circuit.

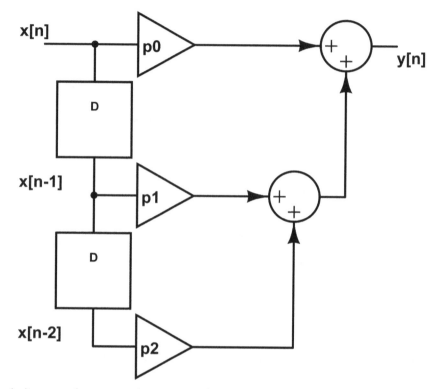

Figure 6-16: Block diagram of a generic discrete-time finite impulse response (FIR) filter. The "D" blocks are delay elements and the triangles represent multipliers.

This is a standard DT filter block diagram. We use the triangles to denote multipliers and the "D" blocks to denote delay elements. The delay element delays its input by one clock cycle. Therefore if its input is $x[n]$ then its output is the sample at the previous time step, namely $x[n-1]$. If we follow the data path we find the "difference equation" is

$$y[n] = p_0 x[n] + p_1 x[n-1] + p_2 x[n-2] \qquad [6\text{-}13]$$

This is roughly equivalent to the differential equation for a continuous time analog circuit. Note that, in general, a filter can have any number of "taps." The example above has three taps.

Example 6-4
FIR filters are best introduced through an example. Consider the block diagram shown in Figure 6-17.

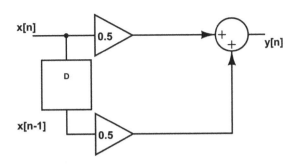

Figure 6-17: Block diagram of an example FIR filter with two filter taps.

In this case the difference equation is $y[n] = 0.5x[n] + 0.5x[n-1]$. Let's start by finding out the impulse response of this filter. To do so, let's start by creating an impulse and then tracing its path through the circuit. Our impulse will be

$$x[n] = [1\ 0\ 0\ 0\ 0\ 0\ ...]$$

We solve for $y[n]$ by tracing the numbers as they go through the system, as shown in Table 6-2.

The impulse response, which by definition we call $h[n]$, to be consistent with continuous time impulse responses $h(t)$, is $h[n] = [0.5\ 0.5\ 0\ 0\ 0\ ...]$. This is convenient! The impulse response is just the same as the filter taps. Also note that the filter response is *finite* in that it doesn't last forever. Therefore, it is a finite impulse response (FIR) filter.

Just as with continuous time signals, we can find the filter's frequency response (the transfer function) by taking the Fourier Transform of $h[n]$.

$$H(\Omega) = \sum_{n} x[n] e^{-j\Omega n}$$

For our example, we get

$$H(\Omega) = 0.5 + 0.5e^{-j\Omega}$$

As with any complex-valued Fourier Transform, we can use Matlab to plot the result. A more practical solution is just to use Matlab to calculate the Fourier Transform automatically. This can be done with the `freqz` command.

```
freqz([0.5 0.5])
```

The resulting plot (Figure 6-18) indicates that our example is a low-pass filter: frequencies near $\Omega = 0$ rads/sample are passed with a gain at or close to 1. Frequencies near $\Omega = \pi$ rads/sample are rejected with gain at or close to zero.

Table 6-2: Impulse response for the system shown in Figure 6-17.

Sample(n)	$x[n]$	$x[n-1]$	$y[n]$
0	1	–	$0.5 \times 1 = 0.5$
1	0	1	$0.5 \times 0 + 0.5 \times 1 = 0.5$
2	0	0	$0.5 \times 0 + 0.5 \times 0 = 0$
3	0	0	$0.5 \times 0 + 0.5 \times 0 = 0$
4	0	0	$0.5 \times 0 + 0.5 \times 0 = 0$

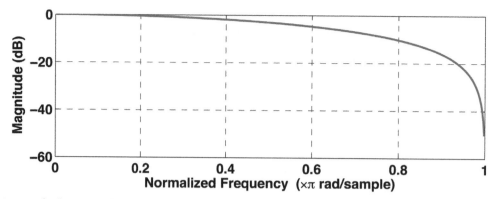

Figure 6-18: Transfer function of FIR filter $h[n] = [0.5\ 0.5]$. This is a low-pass filter. Low frequencies are passed with gain 0 dB (equivalent to a gain of 1) and high frequencies are rejected with a large negative dB gain (equivalent to 0).

Finally, let's consider how we can figure out the output $y[n]$ in response to some arbitrary input $x[n]$. As an example, let's let

$$x[n] = [\ 1\ \ 3\ \ (-2)\ \ 4\ \ (-1)\ ]$$

To determine the filter output $y[n]$ we can do one of two things. The first method is to use sum of weighted shifted impulse responses. We can think of our input as a sum of shifted and weighted impulses. Therefore we can calculate the output $y[n]$ as a sum of shifted, weighted impulse responses.

1×0.5	1×0.5				
	3×0.5	3×0.5			
		−2×0.5	−2×0.5		
			4×0.5	4×0.5	
				−1×0.5	−1×0.5
0.5	2	0.5	1	1.5	−0.5

The first row shows the impulse response of [0.5 0.5] scaled by 1 and shifted by 0, in accordance with the first sample of $x[n]$. The second row shows the impulse response scaled by 3 and shifted to the right by 1, in accordance with the second input sample. Summing down the columns, we find that $y[n] = [0.5\ 2\ 0.5\ 1\ 1.5\ (-0.5)]$.

The other method for finding $y[n]$ is to convolve $x[n]$ and $h[n]$. As in continuous time convolution, this involves flipping, shifting, and adding.

n					0.5	0.5					$y[n]$
0	−1	4	−2	3	1						$0.5 \times 1 = 0.5$
1		−1	4	−2	3	1					$0.5 \times 3 + 0.5 \times 1 = 2$
2			−1	4	−2	3	1				$0.5 \times -2 + 0.5 \times 3 = 0.5$
3				−1	4	−2	3	1			$0.5 \times 4 + 0.5 \times -2 = 1$
4					−1	4	−2	3	1		$0.5 \times -1 + 0.5 \times 4 = 1.5$
5						−1	4	−2	3	1	$0.5 \times -1 = -0.5$

The second row shows the input signal flipped and shifted so it lies just under the first sample of $h[n]$. The second row shifts the input to the right by an additional sample. For each row, we multiply the flipped-and-shifted values of $x[n]$ with the corresponding values of $h[n]$ and sum the results. In the end, we get the same answer as before for $y[n]$, as shown in the right-hand column.

Of course we could just use Matlab to accomplish the same goal using the following steps.

```
x = [1 3 -2 4 -1];
h = [0.5 0.5];
y = conv(x,h);
```

Note that the filter $h[n] = [0.5 \quad 0.5]$ is a moving-average filter. It gives us the average of the current and previous samples. Moving-average filters are very convenient for smoothing noisy data to reveal otherwise hidden trends. This assessment agrees with the Fourier Transform show in Figure 6-18; any time you hear the word "smoothing" you should think "low-pass filter."

FIR filters have a number of features that make them a popular tool for engineers. One advantage is that they are relatively simple to design and implement; even a simple two- or three-tap FIR filter can be very useful in the right circumstances. Another important property of FIR filters is that they are always *stable*. A filter is stable if a finite input always produces a finite output. We always want our filters to be stable in order to avoid outputs that blow up to infinity. Since all FIR filters are by nature stable, they are always a safe bet when designing discrete-time systems. However, while they are stable, FIR filters are also limited in terms of their sophistication. In order to build more desirable filters such as Elliptic or Chebyshev, we must use IIR filters.

Section 6.5 IIR Filters

A second class of discrete-time filters is called infinite impulse response, or IIR. IIR filters are a little more complicated to analyze than FIR filters, but they are also capable of producing better, more sophisticated filters. Figure 6-19 shows a sample IIR filter.

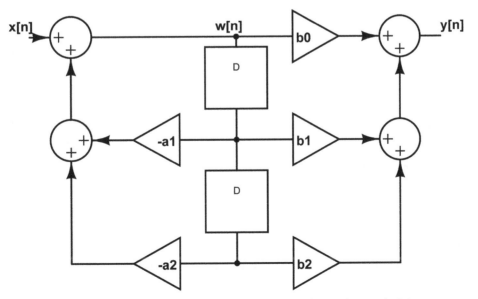

Figure 6-19: Generic three-tap IIR filter. The intermediate nodes have been denoted $w[n]$.

Our first order of business is to understand the relationship between the input $x[n]$ and the output $y[n]$. Note that the IIR filter has an intermediate node that we've labeled $w[n]$. By following the block diagram, it should be reasonably straightforward to see that

$$w[n] = x[n] - a_1 w[n-1] - a_2 w[n-2] \quad \text{or equivalently}$$

$$x[n] = w[n] + a_1 w[n-1] + a_2 w[n-2] \tag{6-14}$$

and

$$y[n] = b_0 w[n] + b_1 w[n-1] + b_2 w[n-2] \tag{6-15}$$

Ultimately, we want to combine these two equations to get a single equation relating $x[n]$ and $y[n]$ that doesn't depend on any $w[n]$ terms (we're not interested in $w[n]$). The process of combining these equations is a bit involved. Suffice it to say that the final answer comes out to be

$$y[n] + a_1 y[n-1] + a_2 y[n-2] = b_0 x[n] + b_1 x[n-1] + b_2 x[n-2] \tag{6-16}$$

We can prove Equation [6-16] by substituting in Equations [6-14] and [6-15] and showing that both sides of the equation are equal. Note that in general, IIR filters can have as many taps (or levels) as needed; Equation [6-16] can be modified accordingly without having to re-derive it from scratch.

Comparing Equation [6-16] with the difference equation of the FIR filter (Equation [6-13]) we see that the main difference between them is that the IIR difference equation is "recursive" in that old values of $y[n]$ are used to calculate new values of $y[n]$. This characteristic of having "memory" of its own output allows the IIR filter to exhibit some terrific properties.

Example 6-5
Consider the IIR filter shown in Figure 6-20.

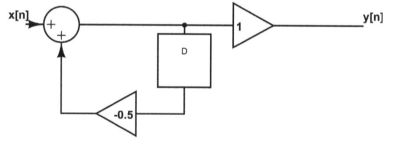

Figure 6-20: Example IIR filter.

We can tell by inspection that the difference equation must be

$$y[n] + 0.5 y[n-1] = x[n] \tag{6-17}$$

Let's determine the impulse response $h[n]$ by applying an impulse to the input and seeing what the output is. We can determine a formula for $y[n]$ by rearranging Equation [6-17] as

$$y[n] = x[n] - 0.5 y[n-1] \tag{6-18}$$

We can now determine $h[n]$ by applying an impulse and solving iteratively.

n	$x[n]$	$y[n-1]$	$y[n]$
0	1	–	1
1	0	1	−0.5
2	0	−0.5	0.25
3	0	0.25	−0.125
4	0	−0.125	0.06125

Notice that, unlike with the FIR filters, the impulse response here doesn't quite reach zero. In fact, although $h[n]$ will get asymptotically closer to zero, it will never actually get to zero, no matter how far out we let n go. Therefore we say the impulse response is infinite. Pretty spiffy!

Our next order of business is to determine what kind of filter this is: low-pass, high-pass, or other. Let's start by employing the low-budget approach. We'll create a low-frequency signal ($\Omega = 0$) and a high-frequency signal ($\Omega = \pi$) and pass them through the filter. Then we'll see which one was passed and which one was rejected and determine what kind of filter we have. We start by creating a low-frequency signal, $x[n] = \cos(\Omega n)$ where $\Omega = 0$ rads/sample. As we know, this is just a constant signal. For our example, let's use ten samples.

n	0	1	2	3	4	5	6	7	8	9
$x[n]$	1	1	1	1	1	1	1	1	1	1

Using Equation [6-18] as our guide, we can solve for $y[n]$ iteratively as follows.

n	$x[n]$	$x[n-1]$	$y[n]$
0	1	–	1
1	1	1	0.50
1	1	0.50	0.75
3	1	0.75	0.62
4	1	0.62	0.69
5	1	0.69	0.66
6	1	0.66	0.67
7	1	0.67	0.66
8	1	0.66	0.67
9	1	0.67	0.67

It appears that, after an initial settling period, the output converges to a constant stream of about 0.66. Our gain at $\Omega = 0$ is therefore $0.66 = 20\log_{10} 0.66 = -3.52$ dB.

Next, we repeat the process for $\Omega = \pi$ rads/sec. In this case, our input signal is $x[n] = [1\ (-1)\ 1\ (-1)\ 1\ (-1)\ 1\ (-1)\ 1\ (-1)]$. Using the difference equation yet again, we see that our output signal is $y[n] = [1.00\ (-1.50)\ 1.75\ (-1.88)\ 1.94\ (-1.97)\ 1.98\ (-1.99)\ 2.00\ (-2.00)]$. Again, after an initial settling period, the output settles to about double the value of the input for a gain of $2.0 = \log_{10}(2.0) = 6.02$ dB.

Considering that we had negative gain at low frequencies and positive gain at high frequencies, it appears that our filter is a high-pass filter. This is a great finding, but we really had to expend a lot of effort to figure that out. Let's see if there is a faster way.

When we were dealing with FIR filters, computing the transfer function $H(\Omega)$ was relatively straightforward—all we had to do was take the DTFT of the impulse response $h(n)$. However in the IIR case, this approach is complicated by the fact that $h(n)$ is infinitely long. So while we could technically write

$$H(\Omega) = 1e^{-j\Omega 0} - 0.5e^{-j\Omega 1} + 0.25e^{-j\Omega 2} - 0.125e^{-j\Omega 3} + \dots \qquad [6\text{-}19]$$

the problem is that the expression would go on forever and we'd have no way of precisely evaluating it at a particular frequency, such as $\Omega = 0$ rads/sample. Note that this approach is not wrong—Equation [6-19] is technically correct! It's just that it's not practical to work with since it doesn't give us a closed-form expression for $H(\Omega)$.

It turns out that the best approach is to copy something we did with continuous-time signals. Recall from Section 4.5.1 that in order to solve for $H(j\omega)$, we took the Fourier Transform of the differential equation. Let's repeat that process now for the difference equation in Equation [6-17]. We can say that the DTFT of $y[n]$ is $Y(\Omega)$, but what is the DTFT of $y[n-1]$? This can be addressed by reconsidering Equation [6-9] as

$$\begin{aligned} \text{DTFT}\{y[n-1]\} &= \sum_n y[n-1]e^{-j\Omega n} \\ &= \sum_m y[m]e^{-j\Omega(m+1)} \\ &= e^{-j\Omega}\sum_m y[m]e^{-j\Omega m} \\ &= e^{-j\Omega}Y(\Omega) \end{aligned} \qquad [6\text{-}20]$$

Armed with this useful information, we take the DTFT of Equation [6-17] and arrive at

$$Y(\Omega) + 0.5e^{-j\Omega}Y(\Omega) = X(\Omega)$$

$$Y(\Omega)\left[1 + 0.5e^{-j\Omega}\right] = X(\Omega)$$

$$H(\Omega) = \frac{Y(\Omega)}{X(\Omega)} = \frac{1}{1 + 0.5e^{-j\Omega}} \qquad [6\text{-}21]$$

Equation [6-21] gives the transfer function for our example IIR filter. Let's see if it predicts the cosine responses we found earlier solving the difference equation. If we substitute $\Omega = 0$ rads/sample into Equation [6-21] we find $H(\Omega = 0) = 1/(1 + 0.5) = 0.66 = -3.52$ dB, exactly as we did before! If we try $\Omega = \pi$ rads/sample, we again get $H(\Omega = \pi) = 1/(1 - 0.5) = 2 = 6.02$ dB.

The general form of Equation [6-21] for a generic IIR filter is given by

$$H(\Omega) = \frac{b_0 + b_1 e^{-j\Omega} + b_2 e^{-j\Omega 2} + \dots}{1 + a_1 e^{-j\Omega} + a_2 e^{-j\Omega 2} + \dots} \qquad [6\text{-}22]$$

As a final note pertaining to discrete-time signal processing, we'll take a closer look at Equation [6-22]. This equation is the Fourier Transform of the system-difference equation. However, as we learned in Section 4.8.1, we have another tool at our disposal called the Laplace Transform. In the continuous-time domain, the Laplace Transform allowed complex frequencies of the form e^{-st} where $s = \sigma + jw$. We noted that if we let $\sigma = 0$, the Laplace Transform would become the Fourier Transform.

As it turns out, the Laplace Transform also exists in the discrete-time domain. It is typically referred to as the z-transform and it uses a discrete-time frequency $z = re^{j\Omega}$. As before, the discrete-time Fourier Transform is a subset of the Laplace Transform; if we limit z to the unit circle by forcing $r = 1$, the z-transform reduces to the DTFT. With this in mind, Equation [6-22] can be re-expressed as

$$H(z) = \frac{b_0 + b_1 z^{-1} + b_2 z^{-2} + \dots}{1 + a_1 z^{-1} + a_2 z^{-2} + \dots}$$

[6-23]

In this context, values of z that set the numerator equal to zero are called "zeros" and values of z that set the denominator equal to zero are called "poles." As was the case with continuous-time signals, circuit designers typically start by specifying a filter type and order, and the design software tells them where to place the poles and zeros. Once the poles and zeros are specified, the values of a and b can be determined and the discrete-time circuit can be constructed and tested.

We stated earlier that one of the advantages of FIR filters is that they are guaranteed to be stable. IIR filters are a bit trickier to work with in that it is entirely possible to accidentally build an IIR filter that is not stable. An unstable filter may produce an infinitely large (unbounded) output in response to a bounded input; this is generally not a desirable quality. Generally speaking, an IIR filter can be kept stable by requiring that all the poles be located inside the unit circle (i.e., magnitude less than one). This is analogous to our requirement that stable continuous time filters have poles only in the left-hand plane.

Section 6.6 Summary

Continuous-time signals can be sampled to create discrete-time signals. Using a sampling rate of F_s samples per second, a continuous-time frequency of $\omega = 2\pi f$ radians per seconds becomes $\Omega = 2\pi f / F_s$ samples per second. Although Ω can technically take any value, DT frequencies are unique only on the range $0 < \Omega \leq \pi$ radians per sample.

A continuous-time signal with bandwidth B Hz must be sampled at a minimum of $F_s = 2B$ Hz in order to avoid aliasing. Aliased signals are undersampled and will appear as lower frequency signals than intended.

The frequency content of DT signals can be determined with the discrete-time Fourier Transform, which states

$$X(\Omega) = \sum_{n=0}^{\infty} x[n] e^{-j\Omega n}$$

The DTFT is always periodic with period 2π radians per sample.

There are two main types of discrete-time filters: infinite impulse response and finite impulse response. FIR filters don't have any poles and are always stable. IIR filters will be stable as long as the poles lie within the unit circle.

Every piece of the discrete-time signal processing pathway corresponds to an equivalent aspect in continuous time. The block diagram (Figure 6-16 and Figure 6-19) is equivalent to an analog circuit. The difference equation is the discrete-time equivalent of the differential equation. Just like an analog system, discrete-time systems have an impulse response whose Fourier Transform is the transfer function that tells the frequency properties of the system. Finally, the system output can also be calculated directly in the frequency domain by applying

$$Y(\Omega) = H(\Omega)X(\Omega) \qquad\qquad [6\text{-}24]$$

CPSIA information can be obtained
at www.ICGtesting.com
Printed in the USA
LVHW060532111121
702998LV00005B/25